ISBN-13:

978-1492746775

ISBN-10:

1492746770

The End of The Universe: Universal Verses One: Stellation

Universal Verses1: in three-parts: Stellation

1. Stellation...

2. The Universal-Axial.

3. Universal-Axial Stellation.

1. At first there was neither being nor non-being.
There was not air nor yet sky beyond.
What was its wrapping? Where? In whose protection?
Was water there, unfathomable and deep?

2. There was no death then, nor yet deathlessness;
of night or day there was not any sign.
The One breathed without breath, by its own impulse.
Other than that was nothing else at all.

3. Darkness was there, all wrapped around in darkness,
And all was water indiscriminate: then
That which was hidden by the Void, that One emerging,
Stirring, through power of Ardor, came to be.

4. In the beginning Love arose,
Which was the primal germ cell of the mind.
The Seers, searching in their hearts with wisdom,
Discovered the connection of Beings with Nonbeing.

Rig Veda (around 1400 BCE India) X, HYMN CXXIX. Creation.

The End of the Universe: Universal Verses1: Stellation: in three-parts:

1. Stellation

2. Universal Axial:

3. *Universal Axial: Stellation.*

1. Stellation: in three-parts:

-Welcome to The-Universe…

-The End of The Universe:

Between nothing and the start of everything else: without so much of a Big Bang!

Or, as a *sumping* deep dull thumping *Thud*! Unseen unheard

But then felt enough to know that something tremendous had occurred.

By the smallest scalar sizing by any minimal metric measure

By which anything can be measured or known

Something *i*ncredibly wonderful: a point of inflection pierced held

Profoundly weighty waited credible instantaneous false-vacuum framing

Zero-arrowing doubling negating neutral-booming force-field vector
Tipping tripped synergy wave-forming foaming:

-The Universe!

Transcendental! Continuous inconclusive radial half-spin up and half-spin down
Full spin circumferential-spheroidal in magnitude-unresolving…
Indeterminate-*terminal* Being: A point in-and-of *seemingly…unendingly…*
Approaching-retreating *perfecting*-infinitude with no net-charge as yet
Assuming tentative-tenuous potential seeking-*halogenic* as if forever-becoming…

-Dia-metric *almost* equal radii-rationing:

-Dia-magnetic *neutral*-repulsion…

-Doubling negating false-*vacuum*…

-*E*lectronic-attracting nucleating gravitational…

-Macro-Wave…

-Microwave length and frequencies…

Of every probability possible squared-area hyper-cuboid voiding…
Hollow-surfacing One to two to three to four dimensional:

-Quadripolar…

Stretching stretched-equilateral angular-extended:

-H*exing enchanting i*maginary…

Almost Solid-block perfectly stationary-spheroidal as yet assumed:

-*S*ingularities'…

-Rotating!

-Universal!

-Stretching! Each of Us! Now! *differentially* constantly energetic-masses' *equatorial*-bulging polar-capped axial irregular-size and shape-shifting dependent on energy-mass and rate of rotation predicting-model existing:

-Universal *dense* Black-Hole…

-Macro-wave Cosmic *background…*

-Neutron-Star *core micro-wave…*foreground pulsing-Quasar-Stars and Galaxies and Atoms free-falling:

-The Universal-Space:

-The Universe!

Forming ejecting material along polar-aligned radiation-axis compressed glowing Event Horizons *lit…*

Hovering on the *shadow*-edge radiation falling-out floating collided-disorder:

-From *purely*-lawful equational-*constant*: Functional Order…

-Dis-Order!

-What-Order!

-Algorythmic...

-Now!

More than all combined entropy

Rotating-particles from shortlived virtual anti-neutral

Double-negative annihilating reducing:

-Event Horizon...

-Proton...positron...

Lifting from and to:

-Points of no-return...

White hot *Light! Flashed!*:

-From Absolute-Zero...

Nothingness-seemingly: Cold-exploded moving...

-Linear *i*nflationary-expansion...
-Massively-heating from the interior...
-Everywhere!

The gaps in-between evaporate freezing radiating fractional-degrees

Above:
-Primordial Black-Hole...

-Neutron-*blank*...
In and around: immediately suddenly from: *Nowhere:*

-Alpha-Beta-Gamma to X-Ray…

-Burst! Black! White! Light! Everywhere!

Bounced-back then slowed almost ceased…as instantly blown-apart again:

-Neutron negative-electron colliding anti-electron positron-

Proton…

-As Proton Lone-Stars and together…

-Galaxies *forming*…

Neutrino-phonon and photon forcefield.

Laser-smooth uniform-equivalence *shattered clouding*

-Evaporating…

-Condensing…

Collapsing crunching cool-clumping:

-From Zero to atomic-*molecular*…

-Compounding:

-From Absolute Zero-Mass…

-To potentially infinitely…

-Finite hot-light and starry-Galaxies spinning-off…

-*We*...

-*Dark*-nucleons holding-off...

-As All Dark-Energy expelled...

-All Mass now impelling compelling Matter...

-As Neutron-Proton Atomic-Molecular...

-Solar-Stars...*positronic*-Planets and electronic-*Moons*...

-Asteroid and cometary...

-Meteoric Force-Field! Each of Us!

-Combining-*colluding*...

-At half the heat to light electro-magnetic energetic-speeds...

-Each necessarily different as distant and different Place and Space and Time...

-Momentum...

-Inward particle dense-region *emptying*-outward...

-Inward collapsing again barring spiral-Galaxies...

-Spinning-off irregular ratio-mass to energetic boson-nucleic...

-Constant-steadying factors amongst the seeming chaos...

-Lone-Stars and Free-Radicals!

-Rough-shod asteroid and rounded proto-Planetary and lunar pieces...

Stilled barely moving if at all impossible yet to tell

The *vast*-distances now *lit* by Our-Selves
Stabilising collapsing as in homogenous irregular-particle:
-Summing as-over-*history*…
-Motion-phase transition *vacuum*-bubbles
-*Imaginary*-numerals as infinite potential-Time…
-Started within our Own finite-edges…
-No boundary-beyond…
-No forward or back symmetry as asymmetrical moving…
-Arrows of Time and Nature. Nature as Time. Time as Nature.
-Each of Us: Being. The Originating *inverse square root unifying-Unit.*

Outer-tangent doubling raddi-quartering diameter
To circumference area...
Natural-logarythmic aerial-aery virtually eventually circling curved around…*almost* returning to The Start.
The Beginning: Each *uniquely*-constant half-life permanent moment.
Varying-finitude sprung-doubling radial-diametric distance-sprung
Open-strings plucked closing-looping
Almost precise repetition played

Immediately drawn:

-Naked-Neutron anti-*Quark negatively-twinned electronic magnet-field…*

-Proton-positive surviving…

-A Quantum Quark for Each of Us.

2.

From-*Nothingness*

Dark-Beginnings perhaps *another* Universe just like-our own:

-White-*light*-Starship radiant...

-Black-gapped...

-*Wrecking*-ball breaking-up as putting-together...

-Our-Selves! Neutral-neural blank-slate...

-Doubly negating neutral-*nothingness*

-Radiating-raised with no net-charge as-yet...

Assuming tentative-*tenuous* halogenic-tended:

-Proton-positronic!

-Outer Electron-negative *switching*-positronic *cloud*-forming force-field *neutrino flying-off...*

-Anti-quark paired in defence...

-Or attack now no-longer *infinite* but finite...

-Massive-*Nothingness...*

Stretching Universal-Space

-Equilateral tri-angular extending...

Almost-equilateral Each of Us emanate continuously recurrent-

repetitive...

Labouring working the-point:

-Originating inverse root-squared unifying-Unit...

Each *continuously:*

-Recurrent of the originating inverse root-squared...

-Outer-tangent *diameter-to-circumference* area...

-Quadruple tri-line body-base cubed pentatonic...

Encircling:

-Neutron-Proton Nucleon...neutrino displaying...

*D*istance-surfacing balling uniquely ubiquitously

*Perennially e*lectron-sprung

I*ngredient*: armed and legged barring spiral-Star Galactic.

Diaphragm and girdle-bone as bafflingly-erected head and shoulders

Arms and legs as hands and feet formless-knowability *in itself* seemingly

Divine-energy...immateriel-*stomaching...*

Ourselves yet dark-hearted eaten or not To Be eaten.

Given taken as in the taking

As if by theft or trapped in-reciprocal kindness.

Traipsing-out:

-Particle microwave-*wave...*

-*D*etectors...

-Detractors...

Not-so-*mysterious...*

-Hydrogenic Each of Us as in Our Own Known Space and Time.

Marching in number holding-out against
Free-travelling electrons de-localized as to unknown
unknowable *Battle*
Collectively a Universal WarZone.

3.

Each of us *constant-complex* altering:

-Nuclear atomic electro-magnetic...

-Intermediary-*Graviton...*

-Monopolist inducting-induced...

-Chain-store...

-Chainsaw...

-Competing-competitive aggressive fast...

-Passive-slowing...

Weakly incomplete electronic-cyclonic
Finitudinal-truncating:

-Number-field...

Integral-spin grouped:

-Each of Us set-out...

Setting-out set-out.

Each to Each immediately:

-Split gravitationald-disc…

Grouped one to every possible few numbers

Leading to innumerable-number…

Tip-ended double-arrowing down down up up down:

-Quark-Nucleon electro-magnetic field as gravitational mass-energy…

Against an impenetrable light wall-boomed…

Golden elliptic-egg shell-seemingly *Thudded!* Darkly energetic-amassed.

Expanding-faster towards the fastest speed of sound

To The Listeners:

-The Measurers…

Be-spoken speaking-spaces in-between

Before-becoming:

-Power and Light!

-The Seeers!

Slowly now colouring in the-background.

Universal-microwaves started *twice fallen tumbled stumbled*

Bounced back-up…thrice

Parsimonious *perhaps*-luckily

Perhaps not.

As a Matter-of-Degree

Pre-planning now…smashed clashing:

-Light-Sabres!

Lightening in any of Every direction

Vacant *but that*-plan already-*gone*:

-In Space and Time…

-Sounded-Light!

Split-second spread! Now positively-turning returning two-thirds spin-charge

Equivocating including Our-Own *negation*:

-Anti-down-quark doubling-up again neuralising negative…

-Positive Quark-Proton…

Each of Us *again and again and again…*

In The-Darkness surmising glue-balling

Re-centring low-mass lying:

- *Gravitational-Gluon…*

Exchanging *nucleonic*-subatomic *strongly*-nucleating

Full-integer spin mass-particle *leaking:*

-*U*nstable!

-Stable?!

Weakened short-lived seconding movement clocking

Phonon-quickened non-existing into heat and light:

-Neutronium switching Positronium…

-Ever?!

Moved across linearly the empty space

Making-up the Universal-Space:

-Changing?

Felt squeaked-cleaned and thoroughly-*centring*

Thudded! again. Open-centring:

-Pi-onium…*electronic…*

Drawing-in and reducing-excess…

Full-spinning-off s*tinging* spinning-*positively* stingy-releasing holding-onto…

Sequential-ascribing…*selecting-out held onto every unique member-number*:

-For-Millennia kept! Each One of Us!

- Of Us!

-To The End of Us!

Each Proton preserved-intact

So far:

-Growing gardening-in: The Universal-Space!

-Quadrillion! Light Years to Now!:

-To the End of the Universe. The-Measurers…

-All!

4.

All-interlinking overlapping fractal-point functional fractional:

-How many?

-One to two to three to four and more dimensional...

-Ten…Twenty-Six even…27…oddly:…

-Deterrence…optimal pay-off matrix…

-3.1415…*equivocating* half-point pioneering *prime-numbering*…

-Equilateral-triangulating turning-frame…

-Triangulating-set of *real* rational-number quotient….

-The common-counting numbers up and down…

-Inward as outward *caught*…

Entangling musical-scale tinkling:

-Positronium…

-Pi-onium…

Fractional four-dimensional moving linear-points.
All at-once turning amongst ourselves
With attraction and repulsion recognizable
Denominating-nominating nominal subtraction
Only adding oddly divisive multiplying progressment unevenly
Heaving sprayed in all directions across The Void:

-Time and Nature…

-Mass and energy…

Begun-together *simultaneously*-separating so fast and now so-slowly:

-Bounce-back...

-Back-bounced in:

-Virtual-Weightlessness...

-Impossible logarithmic-construction

-Weakly electro-*magnetic-fielding:*

Quarks! United! Sink-trapped each of us Zero-Spin *electro-*static bound

Acting from within now upon Each-Other:

-Pi-ons...

Circling and encircling as on the spot

As mouthing or anus-disappeared...Stopped!

Collided-with...colluded somehow *spheroidal* rolling-forward

Flattening-elliptic moving-linearly across *live*-wires:

-Intermediary *imaginary i-fundamental particulating...*

*A*lmost encircling swelling swollen with frightened rage!

Open-ended tubules-spoked *spinning-out* in Universal-Space.

5.

In-*darkness...*

Bi-centring ellipsoid elastic-stretched conforming

Circular-linear disappearing *vanishing*-points retaining

Minimal trigonometric cubic-equational:

-Critical-axial crossing difference-resolving cyclotomic cyclic cubic fields:

-Totally *Real!*

-Polynomial permutation-rings at the same moment-of-Time:

-Never-again to become…

As naturally fallaciously *facetiously…*
Each of Us in motion at the same-*i*nstanton *exchanged.*
Secanting-variating flat-top sloping planar sub-tending
Isosceles tangent sliding-down or upward as sideways inward or outwards IT mattered not. Yet squaring the hyperbola:: The *Universal Parabola created* free-fall parachuted-in. Under the *empty*-product
To the dangling axis of suspension somersaulted
Head over heels acrobatic head-spinning tailing
Pulled in all directions at-Once.
Falling in *any*-direction
As bungee-strung netted elastic-trampolining
Re-bounding almost equi-partition distance-penetration.
Seemingly-colliding yet unevenly inevitably crashed-into
The continuously-pitched-distance inherent pure endless:

-False-*vacuum*...

As yet bounced side-lined de-limiting bulging booming
boomed;
*Thud!*ded re-sounding bounded
Contrarily pitted-against.

Each other deeply-*inspirited seemingly* boundlessly
A-bounding *reverent*-riverine *emptiness*
Inward bounced back-up and of Ourselves…
Started once fallen-into twice returning
In any of every direction but this:
  -Now Positively-nucleate...
  -A*nti-down-quark doubling-up Quark…*
Of this returning two-thirds spin-charge equivocating
Down-again and again and upwards again and again...
Low flying-mass *gravitational-gluon* exchanged stuck-together.
In the darkness surmising glue-balling re-centring:
  -*Nucleonic*-subatomic *neutral-Neutron* strongly
nucleating…
  -Double-Negative Full-integer spin…
  -Mass-particle *leaking* unstable!
  -Neutronium!...short-lived...
Moved across the empty open centre:
  -Pionium-*positronium…*
The final-thirded reduced-excess spun-off.

Stinging-stingy spinning positively-sequential ascribing:

-Uniquely numbering...

-Selected chosen spaces as of those available...

-Each One of Us...

-To All!

-Of Us?

Electron positron-switching stable

String-point mass-recessing precessing

Processing multi-dimensional:

-How many?

-Dimensions?

-Universes...

-As One.

-As All.

6.

Coiling crawling cooling inner-space as re-heating resonance containing
Pneumatic-dynamic drilling-down as in:
-Electro-magnetic winding-down the *proportional variable energy constants of Quanta...*

Each of us *localised little* Gravitational
Closing-oscillate wavelength-frequency fundamentally asymmetric
Regulating super-symmetric super-positing
Radial-spinning *Radian* reaching *searching*-out...

Turning twin-polar stretched elliptic five-point centrifugal
Rebounded-sounded cyclical-square pinned.
Spat-out cuboidal-helices set-square:: pentatonic point-planar light-*potential…*

Each individual-variable arresting and re-starting hyperbolic
Sub-tending pining oblong-obelisk sensed
Each of us as boldly **black** as all the deep-**black**ness all-around
Pre-*reflective*:
    -Golden-section searching number 1.6180339887...
    -*Regular:* Alpha-altering plus and minus...
    -Universal-*Inflaton* naturally logarithmic…
    -*Ever approaching infinity b*ecoming outer-space signal-simple *intelligence* triangulating…irregular-moving direction and momentum…
A sharing-hexing *threat* imminent-simultaneously
From and within and without definite-shape.
Over a seeming Cosmic-*Horizon*
Simple to complex double-concave envelope.
Convex holding outer-tension ballooning:
    -*Two to three to four-state quantum...*
    -*B*inary tertiary quadrating pentatonic hexing *septimal-tritone…*
Sentinel guarding computorial-superconducting system sonic-vortex cylindrical tunnelling:
    -*Proton singular individuating-mass...*
As electromagnetic-decaying:
    -*Anti-neutrino* entropic released...
*A*ltering freely-waving...
    -Ionespherical:...isotopical...
    -Radio-*active...*
    -*Light!*
Collided dis-attached and let loose:...
    -Nucleonic *ion conductive* corded...

-Absorbed and adsorbing...
Conic spinning-out vortex:
-In-the-hole vacating:
-Directional in direction of propagation once-fixed...
-Fixed-once capturing and/or captured:
Self-propagating mostly made-out:
-Solar-*systemic* Planets and moons and asteroid...
Sprung in-springing out of spin-twisting almost returning equal
Attempting assaying-*perfectly nearly* reality
Yet Each of Us *from-rest* queuing non-equilateral tick-tocking...
Each internal click-clocking against Each-Other.
In the-moving at the same-point moved-from
Differently not *in-differently*
Three-dimensional opening fractious and fragile formed.
Fractional-factoring angular-momentum:
-Four-dimensional...
-Universal-Hologrammatical...
-Effects...any number or wind-blowing seated screens moving apart and together...
-Giant Squid-like Octopus!
-Leopardine...
-Tigrous! d*icing spots as tabby-striped:*
-As Stars that are fished: as fishing...
-Circling swashbuckling switching *short*-lived...
-Relative-Spin-*e*lectron-point pool...
-Annihilating-potential tragedy...
-Comedy Romantic-Neutron-proton electron-positronic...
-Mon-Atomic tonic-chorded...
Stated screaming-chaotic unruly-terrifyingly *radiant*...
Strongly then Weaker:
-*e*lectromagnetic point-cloud...
-Nucleate!
-Colliding! Crashing!
-Fusing then...

-Fission-again!
Crossed transmitting-points determined only by the sonic-wave
T*hudded...again*

Booming *cruel*-harsh callous-*stinging cut*:
-With depleting absence of pressure banded all-around...
And within with nothing *else* to compare yet:
-Three-point pseudo-scalar tensor-split...
Restless reckless ticklish-stirring string-point mass
Stretching binding-bound:
-Quark-Quantum Moment!
-Measured!
Each-*perversely:*
-Non-Zero Sum…*distorting each other half-spin up*

Reverse half-spin down *by apparent orders* of *myriad*-Magnitude
Toward the *ever*-present while in circuit:
-Ratio stretched-cyclic oscillating point strength…
Weakening…
-Fractally-functional common-ratio rationing-number…
-Giving *gifting* every and All-*number...*
-Infinite-secondary…
-Against some particular outer-*vacuum* pressed non-stopped...
-Stopped radio-wavelength conjoined gravitational
electrostatic...
-As fast as…almost all…
Furiously roughly-furrowing *foaming* twisting and turning
Tersely truculently each-bridging snapping and spinning:
-Impossible!
-Star-Car!: *switchback-looping known fractal-finite almost repeating...*

-Each of Us becoming become *ambient*-spectral critically over-weighing:

-Protean *ideal real Hydrogen-gas ionizing…*

Clearly in Super-abundance and All-around:

-Supra-HyperSpace!

Within and with The All-Around.

Alluding-eluding elimination *illumination*:

-Stellar-Super Hyper-Universe!

-Universal Proto-star HyperNoval...

-Force-field energetic-*shockwave...*

Each-of-Us seductively salaciously spiraling *almost out of control...*

Unable to hold it *All*-together any-longer in-complete newness

Each of Us *awaiting* no-longer to make-OurSelves *clear-reflecting inwardly first*:

-All MassEnergy...

-Space-Time...

shifting gear:

-Quantum-fuel-electrical breakdown...

-Each photon-potential colliding with another…

-Each-Proton attracting absorbing and reflecting surface…

-*Podium*-proportional...

-As a surface still relative to another…

-Temperature of the frosty dry-surrounding heat…

-Eating engulfing *e*lectron or spat-out...

-Vacuum-bubble ultra-cold neutron flux-fixing:

Phonic:

-Thud!

-Again!

Pre-empted this time again rebounding from the grasping-degrees of edge

Gasping at the outset *instantaneous*-raveling…

*Thud*ded-back again and through the centre-passing around:

-The *only*-possible numerical-proportion...
-Unique to the cube of travelling *impact*-speed...
-To The-Edge…
-The-Centre.
Each of us naturally resonating vibrating string strung-out.
Discretely-*interacting:*
-Integral hexing helical-holding...
Stepped-spiral stacking crenellating protective
Castellating degree of tilt
And as turreted:
-Pro-active...
-You!
Turning built-in *fine-tuning* current point-number stop-starting off the blocks.

Rumbling purring complex non-equilateral indivisible compacted and compressed *breathing re-fueling...*

Confounding wheeling-on-wheeling
Ranking latched and ratcheted
Each One of Us hatched flanking
One against another and amongst Each-Other.
As on *invisible* force-field cogs and gearings each gripped loosely:
-Equalising stabilizing as near to Zero…minimum as maximum coming-around…
-*Hidden*-variables…
-*Barometric atomic-pressure engine-gauging...*
-Using Universal-Laws…
-Creating Universal-Laws…
-Hidden-variables?
-As function's constant algorythms…
-Except at any-moment recorded, and moved-on.
Engaging disparate-desperate hanging-on insecurely onto the next.

Or to be shattered apart into so many uncountable
unrecognizing
But yet momentary spiderous-parts...
Repetitive-casually exactly almost and never again.
Only *seeming* causally retrospective-layering prospective
Timeful-entire predictive with the feeling of *nothing:*
Yet also *more* than All of *This*. With practice more but never
exactly precision
The ability to slow and control.

Material memoried-*recalling*
Predictive-repetitive events *approximately*-restoring:
        -Each for Ourselves in ourselves and finally with and
for Other and All-Else...
        -The Universe!
Confirming affirming and constantly re-confirming our own
existence.

To Ourselves and Each Other. Ever slightly or completely off-
centre
Circumspect *imaginative* memorific:
        -From The-Off!
        -Decision-*making*...
Once having started in practical-motion
To keep-going and Be.

Kept-going...*swarming* anti-particle anti-doting
Prodigiously-spookily *ghostly*-distancing:
        -Pro-genic-*prodigy* maintaining...pro-Life!
        -Of course!! Each of Us *containing re-generating...*
        -Each-Other! Only *Just:* To keep-going and Be kept-
going...
        To be *utilized* in-continuity...
Remembering the previous *briefly* whilst acting-out: *presently*
Presented in advance of the apparent intimacy-and-
immediacy of:

-The-Facts...
-The-Act?
-The Only Other Event: Next!
In perpetual *darkly*-retrospective
Repeated lost-forgotten and continuously only ever again part re-found.

Continuous sustaining-perpetuity
Forever in-the-*present* planning-out:
-The Future...
-Of Present and Past-present...
-Next-*remembering* thought dis-membered....
-The more most likely than least likely exact precise repeat...
Part-lost: *Forever*-finding as an *after*-thought fore-thought anew:
-As if *pre-thought*-out...
-Free-Will?!

Acting in The-Moment:
-The Moment-Lost.
In the constant presence of Everything-*else*...
With practice *nothingness*
Within and without the present
Forever-imbalancing
The-Past all-around...

Unavoidable forming and re-forming positively-powering
Re-powering with the urgency of:
-The Immediate...
A reflective-retrospectively reasoned-action
After The Action to add to the data-base of successfully survived-activity:
-So Far...
-All of Us! As more than all the compound atoms remaining in The Universe Now!

-How Many?
The-facts as ever *shattered*-apart.

Into so many *uncountable*:
-Nuclear-atomic inter-actioning...
-Beyond: The-Horizontal...
-Vertical...
-In-as-Out

Interactions actual and possible all At-Once:
-The most likely to the least probable:
-A hierarchy of objects: of particles...
-Gaseous liquid solid-plasma clouding-clouded Clocks clock-making...
-The Time-Keepers!
-The Measurers! The Rulers!
-Non-credible irrational...
-Rational-Paradox...
Outward:
-Passive-aggressive...
-True!
-Logically...
Aside now the slightest movement timeful:
-Speeded-up *ready-made...*
-Your Own owned owed-made...
-Space slowing as *cooling...*
-Ice-Cold!
-Absolutely...
-Frozen!

Racing measured each-by-each by Each of Us:
- The Measurers!
-Ruling!
-Routine-imagination *reasoning-complete...*
-Measuring dipolar-moment *travelling...*
-Momentous-*movement* somewhere-*else*:

-Each combined *almost* equal and oppositional polar-quartering…
-Alpha-altering plus and minus...
-Monopolist Mono-Tonic *charged anti*-particle...
-Mono-Pod! *Doubling…*

Particle-attracting and annihilating:
-Timeful darkly-watery *energetic...*
Materialising-momentous nucleating:
-Tri-parton extreme tri-Quark and *Hexing*-Quartz *Feldspar…*
-Anti-quark...
-*Quantum pionium-gravitational electro-magnet nucleonic massing...momentary...*
-Inverse-distance diminishing and increasing uni-modal iterative and *re-cursive...*
Searching: electro-static and covalent-attraction and repulsion.
Perennial-permutation permeating permitting mutation:
-Massively low-to-highest temperature...
-At the internal fastest speed of sound...
-And then there was-*Light...*

Abrasive vacuum-gap crossed
Almost linear *neutrino* fired-off…
Into the chargeless-*neutral* surrounding *ghostly* non-stop movement:
-Algorythmically-entangled...
-Algebraic-neutralising...
-Mutable-mathematical vitalising ionisation-energies negative and *positive...*
Each of us transmitting transporting qubit-unit communicating...
probing...
-Instantaneously both there and *not there...*
-Quantum Quark-*teleportation...*

In-movement moved-across and around the outside.
Back-grounding black-body radiation-burning red-hot foggy-*opaque*
All-absorbing:
-Energy and Matter…
-Time and Nature…
-Total-Universe!
-Un-bounding!
-Sacred!!

Binding:
-Gravitational darkening once-more...:
-Universal-pulling...
Pressing-buttons computational four state indiscrete.
Non-discrete at-once:
-*Spooky*-action at a distance...

Apparent outer-edged inwardly-weighted pressed
Each to Each so that it was *as if* standing-still.
Or if moving at all
On all-surface in the opposite-direction
Of that directly-opposite *almost…*
Drawn-toward as well as away from
Blown-apart photo-*luminescent from:*
-The Centre…
-Of It All…
-Equal and opposite. Almost sum-leftover…
Preference-pixelate phase-transforming:
-Polarities between energy and matter...
-Hypernoval dense disc-centring transference...
-As light gases and heavy-liquid *watery*-metal:...
Thudding-movement throughout unavoidable heaving-heavier *flashing-light*:
-The Big Universe! Greatest-Macro-wave!
-BlackStar! Force-field Solar...
-Eclipsed!

-Eclipsing EarthCentres! Us!
-*EarthQuakes*!
-Swarming!! Swarms !!
Reflected rising into the-*surrounding* filling-space...
-*Hypernoval Universal!*
-Between energy and matter *relative* to the originating:
-Zero point white-light golden gone-out...
Spinning-orbital each-approximating spin-*circulating*…
Freed-radical sped-each *little-universe*...
Each preserving momentarily-paused seemingly pin-point
Prime-phasing perturbing:
-Split tri-quark point-mass re-charging...
As defined by each of us stopping and re-starting:
-Degrees-of-separation and annexation...
-Freedom...*inverting*-movement...
Staying-mute muling sighted heard baying felt *saving*-energy:
-Bounded added-in three-dimensional The Universal-Space…
Of willingness numerically measured and ruling
Degree-of Free-wheeling willing:
-Each of us *positively-Protonic*…
-*Black-Body cyclotronic-radiation.*

The numbers non-redundant *redundant*
Set-potentially not-finite low to high density:
-To each orbiting *electron point clouding*…
-Singularity…*edged*…
-Valence and shell-Core *centring*…
-Atomic-Nucleic…
-The-Universe.

7.

Cracking crackling massing-points holding-apart:
-Each now non-identical electron-charge pushing against each other...

-Each and every charged repulsion shared valence crossing orbital-reaction:

-Atomic *donor*-nucleus and *electron*-compounding overall…

Re-balancing from firing off-balance but for one billeted bulleting bulletin:

-Taking pre-emptive precedence over-tipping electron-positron mass-locked away minimal to massively magnified as magnifying single atomic-mass...

-One to one hundred-billion energy particulate.

-Every massive proton atomic mass-ratio to *electron...*

-Toward 99.9% 1/9th of the Proton's Dark-Matter and Energy@...

- Atomic Solar-System…

-Sun and Giant gas and rocky Planets-*watering…*

-Rivering-streams and ponds and lakes and seas…

-Ley-lines forging…

-*You! Your* energy-and-matter *now...*With neutral-negative pairing electron-positron capture
Attracting angular-triangulating…
By *impossibly*-symmetrically paring parity-*promoting*:

-Universal: neutron-proton-electron positronic *plasma...*
Felt-prickling-points deflecting *reflecting*
Absorbing-recording and adsorbing Play-back *bleeding…*

Oscillating particle-reader reading:

-Electro-magnetically rhythmic-switching-metronomal…
Danced balancing-enjoined tagging-along.
Rounding-out and positioning-quarter containing boxed-in
Boom-Boom booming *thud*ding squealing squeaked *screaming!*
Closed pitch-sized as sided secant-netted
Dropped-in and raised stretched held-back and fired-off:

-Annihilating The *first*-Generational *energy*-speeding...

Unimaginably slightful insightful losing energetic-mass re-*neutralising*
From neutral negative-positive static-pressure pulled and pushed
Pulling and pushing with gusto-gushing…

Galoshing-sploshing splashing
Sled-ride in-a *sliding Great-Vacuum:*
    -Each of us...
    -Spoked-secant-joined-up...
Spoked-wheeling multi-lateral linear ray-line *irradiating…*
Re-bounding longer and shorter broadcasting from:
    -Electro-magnetic field-force...
    -Nucleonic strongly and weakly gravitational...
    -Your Universal-Event...
Jagged ripped-Open *torn-apart* :
    -Necessarily asymmetric tri-or-more part…
    -No-more…
    -New!
Unbalancing-movement dropped-in again.
Impressive and compressive-force's thickening:
    -Femto-metric *positive…*
    -Protonium...
Conic-pointed seemingly massively:
    -Magnificent electro-magnetic…
    -Mono-polar…sub-limating…*ionising...neutrino and more fundamental-particles as fundamental-forces…*
    -Each of Us!
    -Solar-Systemic!
    -Universal Black-Hole!
Balancing net-charge unitarily-finite:
    -Exciton: Universal *hot*-energy...*Light!*
    -Selectron-*sparkicle...*
    -Non-Zero! *Excited-ringing light-ray internal-conversion instantly boiling-away:*
    -Double-paired outer-quadrupling hexing-sextupling...

Continuously *inflaton* inflating suddenly as instantaneously:
-From Absolute Zero cold to Universal Absolute *heat...and Light!*
-Max.: T*hree trillion degrees* thereabouts...
-Min!: Quark-fission light f*usion*!
-Each of Us firing-off neutral negative plus-minus positive…
Universal Plasma cloud forming...

Magneto-electrostatic caged crystalline leaved.
Opening petal *flowering* voluminous as a grain of coralline-sand watery droplet rained. Glorious pseudo-pod each our own bubble-swelling swollen-lunged bellied belied-toothed tongue-pulsing…
Conducting charging-numerating holding-outward:
-Inner-spin-force winded and reverse tipended...

Thrown-out and into paths and passages made-ahead.
Exploding destroyed and re-newed accreting rapidly re-turning:
-Angular-extending captive-capturing quantitive:
-Whole-*hexing* integer-spin...
-Whole-*I*nteger-spin...hexing...
Spinning-back together-colluded avoidance and then:
-Colliding-consuming outer-necessarily ideally-empty equally-expansive…
-In all directions squared cubed-balling *cell-bound doubling-nonet*:
-Across Each-Operative *approximating...*
Imagined-centre indicating circular directional:
-Constantly-mathematical multiple-proportionate...
-Approaching finitude-indicating di-pyramidal imaging:...
-Against a symbolic-logarithmic curve-*perfecting number-point...*

Quasi-crystalline cross-barred prismatic sheer *darkly* transparently-superimposing *almost*-Solid…
Gaseous-Liquid solid cooling freezing grid-framing
Repulsing-*nothingness*
-Condensing-Universe…
-Evaporating energetic-impulsion and with equivocal-expression…
-The instant glinting the almost unending *pure*-darkness amassing:
-Each *Universal Black-Star centring starry-White*!
-The Universal-*Vacuum!*
-From Absolute Zero cold...
-With Universal Maximum *Heat...as light!*
The renewed *almost*-blackness:
-The Universal Void…
-*Phonons*-only *unseen bouncing-off thudded squealing screeching!*
-Absorbing photonic-returning faster…Lit! Super-Sonic booming re-*energising*:
-Neutrino passing straight-*through...*
-Super-fast then sluggish *returning*:
-Energy to Mass...
-Mass to *Energy...*

Anyway *cooling* minutely F*lashStar!*
G*reying* after-image: *blue red green yellow...*
Orange-twinkling all-around
Silver-white *blue* separating-out red-distance.
With nowhere else to go *greying*-into utterly reflective de-reflection.
Complete-darkness under our own sheer bright-lit ballroom-lights
As stars risen into an as yet starless-night.

8.

Complex geodesic-light bending gravitational electro-
magnetically
Pressure-*weighted* rotational balancing-support nuclear…
Unclear in *utter*-darkness again inside-collapsing
Slowing de-pressuring and re-pressing *membrane* turning-
uplifted:
    -Over three-gravity forces heavier...
    -Four-five…
Without safety-harness oppositely oppositaI down-turning:
    -Than Nothing?!
    -From-*nothingness...*
Drawn–back inward again weaker low-energy dropped down
upto:
    -Pentangular nodal-complex point-planar…
Spin-dragging flung:
    -Within The-*moment...*
    -The-Event! Of Time and Nature...The-
Same…*unbalanced...*

Mega-dropping dropped-in soundless blindly spatial-silence
unheard…

Panicked iced-*frozen* panned-pained pumiced *unhearing…*
Tottering teetering *screaming*-sloped slid-sliding down
Toppling Observation platform-podium towering:
    -Axial down-swaying...
    -Hold-on!
    -Hold 'em Up!?

Gamboling rounding-back an oddly doubly tripling
quadrupling
Penta-*tonic* unequally elastic-sided:
    -Extreme and mean median...

Snagging-inward nodal-complex point-planar
De-pressurising and repressing-membranic *Thudded!!:*

-Each of Us...we breathed-life into-Ourselves and Each-Other.
Outward Hydroid-polyps uplifted: one two three
Over three gravity forces-heavier...

Crossed and contrary clatterous
Electronic rattling platform rafting-regional:
-Each-to-Each learning from Our-Selves...
-As All Others' *synthesizing:*
-Vacuum-*Machines...*
Made of metal and mineral rock-scraped and scratched together.

Consumed-consuming
Forged in gas and in dark-waters rafting-along
Ourselves now with all other firing-waves
Spread interference patterning-particulate passing-across...
Passing-through crossed and strung-out colliding-points
Cancelling-out the spaces in-between:
-*Pure*-Vacuum tri-point breaking:
-Quantum-emission tri-parted...
-Again.
Inevitably-*asymmetric* radiating-moving:
-Ditrogonal-modular...

Folding polyhedral-topical idealising idealised imaginings:
-Imaging: *I*deal *Real*-gas...
-Primordial-stable monatomic-state.
*G*rey-silver white separating-out with nowhere else to go.
G*reying* into utterly un-reflective complete-darkness amassing:
-The Universal-Vacuum...
As yet unlit Solar-Planetary ellipsoid-rotation.
To reach-back into the di-polar axial-centre
Torn crab-claw grabbing
Spinel-scary *terrifying* yet-thrilling affirming *confirming...*
Each To Be:

-Our-own projective non-zero non-negative valid proof-positive.

I*ncreasingly-Other de-creasingly inevitably-*reducing also:
-Absolutely...*irreducible...*
-*A*s negatively-*reducible...*
-As contradictory-proven *demonstratively* as by:
-A *Universally* Incomplete-*Paradox...*
-Complete.
-Irrational.
-Incomplete...
Parodying troubling-affine non-affine.
*I*ncomparable gouging-out.

Set-stringline boundary-mediating massively-asymmetric
Infinite-potentiality seemingly boundless and unstoppable:
-And so it all ends and is to The-End...

Oppositely mirrorly and sinisterly *ghostly* in the machine.
With The-Machine-handed-down and foot-pedaling
Pumping like a decorative-fairground organ played a winding-down.
Heard and seen *sapient* salient-connected and combinatorial
Requiring post-rational reasoning-activity to work.
As *imaginative*-toys or Weapons of Mass-Destruction.
Of Life. Potential *pile- up Crash!ing!*
Braking-arresting pushing-forward
Into custody-detained taken-hold of
Risk-factors apprehending-turning wounded-dropping...
Rising colliding force-field breast-feeding pount-pointing
Counting fed-breathing wont-mounded.
Exploding-mountainous and imploding-bubbles of sound and vision.
Fading-*shadow* elongating through vortex-tubular empty-centring
Surrounding *pure*-starlight points-all-around:
-How many?

-Varying Intel...
Attracting and detracting forces' measured to *robotic*-accuracy
Learning-mistakes with double-jeopardy triple and more-adjustment
Next time if there was a next time...
And there was.

9.

Horizontal transverse *latitudes…*
As if defining divining the whole: *leveling-off*
Joined verticular vermicular ventricular.
Upright-standing perpendicular-spinel
Nervous armed radial *rounding-out…*
Unifying *non-trivial* integrating complex-looped again:
-Each numerical-bias leaning...
-Twisting and turning rattling loosely bolted helical screw threaded. Dexstral and sinistral-facing circuital core-headed and tail-surfacing.
Completed full-circuit spin stepped:
-Sonic-conic parallel-helical convergent.
Twisting-end wrapping and sweet-enwrapped. Regulating-regulated switching-circuitry crossed latticed hatched *thatched* through....
Between tri-point podded.
Hyperbole collapsing-on-sides all-around
Cross-string barred empty-*nucleate…*
Falling weighted-down cartoon canton-centring
Non-reflective mirroring-inside.
Positively-charged again nucleo-tidal:
-Quantum-base...

Di-polar super-positioning
Entangling base-pairs ellipsed...
Each Other eclipsing-eclipsed
Tri-point zipping zipped...
Pedalling and contra-pedalling cycloid swarming...

Lasar-photonic conceiving raised-*momentary*…
Each individual dark-crib caged encountering
Spinning-together or counter-spinning…

Opposital The-Same *almost* or *spooky* extreme-opposite
Composite to the outer-*emptiness*:
  -Rotational…
  -Operator...
Inner-mass weighting interpreting quantifying:
  -De-coherent three-body point *quarter*-Powering...
Re-pressurising neutral easy-stopping and unequally re-starting:
  -Atomic-Universal…
  -And so It All ends…
  -And Is. To: The End.

10.

Each-individuating mathematical reference frame-dependent
Rounding out and down shaping empty-husked hulling seated
clamped-in prism *flash*-points locked-in. Looping linear defying curving ovoid spheroidal stretched expansive *increased*-felt gravity-force rounding rationalising down floor up to roof. Around the *evident* walls truncating digital de-limiting:
  -Fractional *real*-number...
  -As opposed to unreal?
  -False?
  -Then?

Cross-sectioning quadrilateral emptying empty-husk.
Whole-section dimensional complex-chaotic rapidly topping and bottoming-Out:
  -Swing-Ship!…*set in-motion*
Setting-on *course*:
  -Switch-blade back-known procedural-*recurrent*...

-Problematic...
-Caused ragged booming-complex...
-Rapidly successfully-repeating *almost*...
Altering memorised and materially surviving comparitive-potential
Extra-low to extra-high particulate-point:
-Leaking highest-temperature...
-To lowest heat-trapped absenting-*cooling...*
-Evaporating condensing to *electron* capture and energy-release...
Burst electro-static spluttered-*radiate*...
Dragging-friction pushed-out again drawn-infinitely many:
-By nuclear-Fission...
-Alpha-decay...
-Fusion to Stay-in:
-The Game!
-Beta optimal radio-*activity*...
-Neutron-Proton electron-phonon...*photon*...
Clustering-decay cosmic-rays:
-Potential-spinning spiraling-Orbital:
Slightly-longer and shorter-delimiting:
-From inside-outside each truncating fractioning-*fractal*...
-Re-numerator *hexing* six-star point-shaping square-root inverse tangential *voiding*...
-Defined *defining*-Self...
-By-Each *inverse*-distancing *blank*-centring...
-Unitary-edged...doubling-crossed diametric-distanced...
-Centre-Axial *compounding* triplicating double-hexing diamontine...
Gushing! Flashing! glinting:
-Point-positive sprung Solar Star-heated cooling...
-Octagonal point-particle gauging...
-The Measurers!

Circuital-grasping started non-stopping button pressing steering around...

-The-Rulers of Us All!

-All of Us!

Surrounded-curved as carving-out…

Of the unending *pure*-darkness *exhilarating* unruly-*zooming*:

-Di-polar elliptic nucleating-ovoid...

-Four-dimensional observation platform-shifting:

Avoiding and voiding combinatorial...

-Regular and more and more irregular regulating-substrate...

-*Greater and smaller* rhombic-palpation palliation....

-*By* nuclear-fission alpha-decay cluster-decay fused Cosmic-Ray...

Slightly longer and shorter delimiting point positive-sprung

Never The-Same Each and Every-centring

Each truncating fractioning fractal de-numerator…

Eight-star-point shaping crab-claw inclining:

-Axial-compounding...

-Digital distal-distancing...

-Opposital...

Octagonal-nonet circuital-grasping re-doubling:

-Rhombicuboctahadronal pressed-into-service...

Steered around the-outside surrounding curving as carving-out of the un-ending *pure*-darkness...*exhilarating* unruly zooming each our-own observation-platform=Domed-window viewing-plate *shifting:*

-Maximum electric-magneto spherical-permittivity in *pure*-vacuum…

Gradually glowing red-to-blue

The earliest most distant and local:

-*Stellation:* pure-energy per unit coronal-spiked radiating-freely-*forming:* proton-electron neutron-positron:...

-Each of us...*strongly-nucleating to* electro-weak magnetic emitting photo-electric *potential...*

11.

-Accelerating particle tailing Cosmic-*Wind...*
Winded popped-*thud*ding...hissing fizzing phishing *buzzing...*
Squealing-crossing transmitting-points' crossed and contrary-clatterous:

-Across *pure*-vacuum tri-point breaking-Quantum *emission...*

-*Inevitably*-asymmetric point's-combined then-folding...*radiating...*

-Ditrogonal-modular folding-polyhedral...

-Typically idealising-toppling The-*Idealised...imagining...*

-Topical-*imaging...*

-Ideal: Real-Gas.

-Primordial: *fractionally stable-MonAtomic state...*

-Electronic rattling-platform regional-rafting...

-On Our Way Back Home!

De-stabilising re-stabilising-directional sustaining machination.
Pushing-forward braking arresting turning dropping rising colliding force-fields
Exploded and imploding circular-spinning vortex-emptying-tubular...

Closing-*centred* surrounding *pure*-starlight points
All-around. Each of Us a firing-wave spreading-out interference-patterning pointing!:

-Particle-colliding points sending yet more wave-lapping top and bottom-feeder *cancelling-out...*

-Watch-out!

Be-spoke speaking-spaces in-between:

-With a Probability-Wave function in Three-Dimensional Space...

-Frequency and Wavelength determinacy...

-Indeterminate.

-Geometrically-multiplying wave-points outwards and around...

-Collapsing at each-point moved moving-through *uncertainty...*

Still. Changing-direction and velocity-speeding accelerating and de-*accelerating:*

-*Constant*-charged particle-collisions...

-Wavelength-frequency...

-*P*honon-particle striking-out photon...

Electro-magnetic gravitational-nucleic:

-Rogue and *freak-wave* space-stretched by a *factor...*

-And some multiplying and dividing...

Immediately cooling solidifying-almost solid liquid-gas:

-Quantum Quark:...

-Giant Gluon Plasma virtual swimming viral-Zoo!

-*M*assless-energetic forming mass-holding moving massively tentatively around:

-We forming *formed...*

-Re-forming re-formed re-forming *every*-Moment!

-Dead or Alive?!

12.

-Dead or Alive?

Each of us circling each and every other:

-Differentiable: each *nucleon energetic-mass...*

Incoherently uncertainly pointy-bubbly-forming

Chain-rippling frothing

Foaming-gas cloudy-lumped:

-Positively-negatively neutral-balancing...again-*almost...*

-And again.

-Un-balanced...

-Kinetic-energetic:...

Working-rising and falling uniquely invariant-variant:
        -Numerating and re-numerating exterior faceted and internal the same...
        -Almost...more or less safe...sealed-in...
        -Almost equally-weighted photo-spherically...
        -*Plasma...*

Oily-dropped motion we each cloud-chambered potentiality
Each of Us in part to-the-outside Universe
To Each-Other.
Each of Us *emulsifying*:
        -Into-being...
Separating-off:
        -Similar...*different...*

Expanding-Oceanic Seas and dust-Desert-watered.
Soapy-bubbles and baubles compact-constricting
By choice de-limiting every *subsequent*-choice
Until there is none-*remaining...*

We spread like disconnecting-daughtering
Sisterly taking-in brotherly familial-fecund bundling
Incredibly slowly as if nothing had changed.
Almost *static*-sputtering phonic photonic-chromatic:
        -Parenting-finite *numeron* nium-bead-counting...
Counted-out:
        -Daughtering-*generational...*
        -Bean's jumping nudged and nudging-along...
        -Brother! Son! Far-fighters and born faraway...
        -In-*piracy* murdered-stolen away...
        -By guile brilliance and versatility...
        -In All-*directions* minimal-fractal *fractional*-sum...
        -Of each infinite-series trigometric-*expression...*

Bounce-back vacating again
Arriving directional fixing-in:

-Direction of propagation uniquely momentarily paused…

-Self and Other propagating: W-X and –Y-ray photon electron-positonic...

Absorbing and now reflecting wavelength:
-Ultra-violet blue to Infra-red...
Radio-static heat and light Big-Wheeling!:
-Big-Wheel!
Freezing outer-condensate crushing-exchange tunnel:
-A1...2...3-*Hydrogen*...Three…
-Helium...
paused…organic neon lit:
-From Absolute-Zero...
Bouncing slowly neutron against positron at the *coldest*:
-The *faintest*-magnetism now weakly annihilating against *normal* matter...
-Burst aneurism guts-burst!
-Gotten!
Each of us initial ground-state resumed-briefly intermediary modeling the whole more and less:
-*Quantum weirdness* the illusion of wholeness within...
-Colder unburned...
-This Time...
-Never-again!
Booming the pitching darkness instantaneously
Both there and not there
In movement moved-across…
And around the outside *instantaneously* both there and not there:
-In-Movement...
-There and *gone*...

Moved-across and around The Outside
Shared altering-angular rectifying-unit switching on-off.

Burned-across redundant equally massively significant-numerical *cancelling-out:*

-But Two into One!

At the inner and outer edge:

-Cosmic universal speed limit:...

-The Great Escape?

-The Great Expanse…

-*This*-Universe...

-Universal outer-curved space...

-Inner curved to the empty radiating centres…

-Collapsed Dark-Star black hole within and all around *inhabiting…*

-The Same in All-Directions…

-Except Each *uniquely* now-distributing background *radiation atomic code colour-patched…*

-Pitched…

Moving Each of Us

Mass-bending *differently* emplaced emplacing-Self:

-Inconstant constants...unequal equalities…

-*Within: The Laws of Nature! dark energy* and *matter...*

And at some *seeming*-Vertex.

Crashed-into and as if *thought-through...*

Leapt into the dark drawn-in….

-Quantifying quantum *fluctuant:...*

-How many?

-Quarks? Quadrillion. Quark-*point* and chain-ringing *strings...*

-Atomic Ionespherical-*Monatom designer-pattern patenting parenting…*

-Met-Atom! *Nucleonic protean hydrogen liquid-gas:*

Practising orbital-capture and release:

-Freed-*electron* and Each Free-Proton…

13.

-Fundamental-*particles* too-numerous to count…

-Odd and *even*-numbers…

-Since every atom once-was and still *is* seemingly...
-Indefinitely-stable entropy-freeing Protean-*electric*:...
-A number at least one more per every ten-billion...
-As there are now Protean remaining...
-One amongst quadrillion perhaps...at least...
-At least...

Equally-equational *gravitationally* separated electro-magnetic
Variously moving unequally together and apart:
-Binary...at least...*equaling...*
-Equational-*nothing!*
-Where All are Equal!
-In Death?!
To *electron*-capture and energy-release combining:
-Each number-Power....
-Killed-paternal fraternal killer unknowingly unfriendly-realizing...
-Out of The Sea...
-Poison of The-Ray-*fish...*
-No gentle death in sleek old age...
-But Now!
-Not 3-clicks old!
Whirlpool-drowning purgatory rumoured false-counsel:
-Not muling-Circe...
-Or Penelope.
Outer-electron likeness repelling *apparent*-other holding-apart:
-Powering-up paired positive and negative *half*-spins...
-Whole-spun Unitary...
Containing continuously maintaining internal seeming at least temporal *naturalistic*-consistency...
-Never-both...
-At the same Time...
-Or Naturally...
-Of neutrino-photonic non-decaying massless *pure*-energy electron atomic-pitched...molecules...
-All the atoms of The-Body...

-Personal Galactic coronal-circulating spiked-*radiating…*
-*All-Together!*
-*Seperating…*
Rounding-out noisily *freely-forming*:
-Proton-electron neutron-*positron...*
-Strong electro-weak magnetic emitting…
-Photo-electric *potential...captured* gained…
-Accelerating particle-tailing negative-poles…
-Cosmic-*Wind...*
-*A metric single-atomic space between nothingness…*
-Splattering selective-quantum fluctuant *ion*-wavelength frequency-compression...

Enjoined around some apparent equatorial-*Halo*:
-Tagging-along anti-genic *determining...*
Doing: Being done-to:
-*Oscillating*-particles' electro-magnetically *rhythmic…*
-Switching metronomal-balancing…
-Causing-clausifying categorising-subjectively objectively-*Doing to...*
Effectively-*affectively* combining added-to *electronic* condensate *misting* inner-smooth manifold
Manifest electric-barrier potential-charging battery-defending
Volting and re-volting turning and spinning…
Angular-extended sped-by and *through*-amount:
-Smaller positive overwhelming negative velocity-voltage: wxyz…
-Compounding half-spin electron neutral-charge phasing...WXYZ…
Flexing outer-focusing shape-shifting
At-least to control the coming and on-going:
-Telescopic Micro-cosmic…
-ABCD…
-Taking some-part some-control energy taking some-charge of matters…

-DCBA…

-Each of us avoiding by at least the equal distance between us:

-More as less...

-Less as more…massless tri-rings…

-Naturally unequally-*sized…*

-Non-circular ellipses...*interlinking*…knotted-braiding…

-String-like charge massed drifting-away from each other repulsing networks:

-Massive heat and *light* fuel –expulsion powering…

-E*mpowering...*

-U*niversal hearing and seeing felt briefly Universal-Cosmic speed-limiting...*

-Temperature equaling atomic-number releasing heat conversion...

-To light...

Scorching immediately-by:

-Hot-*felt speeding*-past as catching-up darkly-*faster…*

-Heard as Seen…

As We Each were immediately slowed
Re-refrigerated as Each to Each-Other
Also:

-Dark mass-moving together…

-Darkly-energetic *expanding...*

Each other similar charged together resisted passed sometimes faster sometimes slower…
Now Each mutually-attracting opposite-charge negative-positive
Forming crystaline-*latticework*:

-Ion valence-bonding molecular-compound…

Radiating-*energy absorbing*-Masses:

-Adsorbing...felt *Cosmic Wind:* First-Generational*: Universe de-generating* entropic-B*lasted! Blasting-out-in…in…out*...again.

Flung-out and around three-corner-hat pocketed tipped.

Chipped ejected-rejecting tipped-negating:
-*E*lectron neutrino-fired...
Sheer shaved-off electron-positronic photon-field potential:
-Polar-disengaged electro-magnetic force-field...
Screen-*neutral*:
-Neutron-neutrino firing off in all directions...
-*Almost*-linear inflating-points passing almost straight-*through...*
-*Choose!*
In All-*directions* to Each-Other:
-*Negating* potential-gravity energy-oppositial:
-*I*onising...
Screen absorbing as reflecting:
-*P-cation* electronic source to N cathode-P-anode...
-*e*lectrode descended terminal primary *electrolytic*-cell...
-*E*lectronic mass-moving power charging *battery*-cell:
-Proton-*electron source* ray-tube...

Stream-*screaming* anion de-termining determinating dementing:
-Universal-flow current electro-magnetic light potential…
-Up-anion attracted to p-anode...*neutral…*
-Nothingness double minus-down…down-again...up…
-Up-again...again…pure-Proton *positive.*

14.
Ever further apart bolstering intra-acting two point resonating as pixilated screenings:
-Polarising elliptic-inverse universe quadratic-*equational...*
-Calculating differential-frequencies between-fields...
-Circulating gouging-out the space *Laser*-Beam…
-X-Ray…
-Y?

-How?
-Why?

Tri-partite infinite constancy-finite displacement
Positively-negating neutralising the expansive now *impure*-vacuum:
-*Us!* The Universal-Space scene-making *characterization...*
-2nd Generational *particle mass-hierarchy...*
-Weighting ***mass**-boson* moving anti-electron positron neutrino-progressing...
-Sonic-phonic photonic-potential colliding...
Almost-perfectly circular:
-Photon colliding with the centre-axial quarter-polarities switched-back:
-Negative squared-co-efficient wave-frequency...
-Elemental positron-colliding centre-rounding...
Pinheaded dowel rod-fired flung-out.
Now only almost *pure* vacuum-darkness
Now only *almost*-perfectly *linear-radiant*:
-*Almost* 4-Billion...
-3 Trillion *clicks* stepped at Once per-second *per-second...*
-Scalar-speed and velocity...
-Direction of magnitude of directional-distance:...
-Three-dimensional asymmetric-hierarchy electron anti-electron firing:...
-Positronic-*photonic...*
-From absolutely freezing minus-300 degrees of heat maxed-out Speed-of-Light...
-To three-*trillion* degrees possibly *returning* neutron proton electron-bullet! *Protonium* short-lived Neutron Decay...*neutrino...*
-Kaon-*strangeness...ghostly...*
-Eta...

Shifting-chaotic

Strangely-coherently formed.
Forming incoherent absurd-apogee:

-Proton collision switch-proton neutron-up up down producing emitting...

-W-boson turned down-quark up down quark proton W-dies...

-Split positron and electron neutrino transparency...
-Entropy-everything reduced to *lighter*-particles...
-*Weak*-nuclear electron-superposited...
-W-boson integer-spin decay into neutrino necessary:...
-Six-Quark:...
-Lepton-Muon-Tau *gluon*-gravitational...
-Weak-*nuclear* force: +/- and:...
-Tau *e*-neutrino bosonic-forces full-integer...
-Electro-magnetic...
-Exchange-interaction...
-More or less...
-More...not *less*...
-Contact-force...

Across a *pure*-vacuum *dirty* without range *strangely defective* radioactive-decay:

-Alpha-Emission Beta-omission of *positronic...Gamma-Rays...*

-Cosmic-n*euron*-overpowering electro-magnetic strongly-nuclear entrapping-*intact...*

-Stronger...
-The *Illusion* of Solidity...
-Altering clouding and de-re-clouding...
-Captured double or triple-Proton *e*lectron circulate...
-Combined-fission fusing...
-Escaped! *Free* proton *and* electron...
-Freely in Space...
-Each repelling held-together and apart...
-Inwardly and outwardly...

But not for long:

-3rd Generational Atomic…
Caught-as cattle or swine…
-Each *exclusive* Quantum Orbital unique position quantifiable...
-Massing anti-gravitational point-orbital static-crackling cackling stated-started:
-Heard Seen *Felt...*

15.
Unevenly smooth wrinkle-aged edged
Ruffled muffled-yells suppressed dumbed-down
Emptied towards the centring...
Along retracting closing-in axes:
-Universal-draining...
-Routed...
-Re-routing...
Faltering haltered-halting *silenced…*
Ceasing-stilled action *pausing...*
-Minimal-maximal…
-At All-points…efficiency…

Gun-trained. Open and acknowledged acquisition and theft
Between working-parts' mental-states not independent
But *thoroughly*-dependent
On axial-tilt
Having All and *nothingness*
At Once.

Doing On Our Own
Substantially-together *alone and apart*:
-Computorial *electronic*-digital ...
-*Spirited-reasoning...* reasoning-*spirit...desirous…*
-Whole-Integer numerator-ratio decretion…*accretion...*
-Prime-irrational and rational…
-Doing and thought-out…
-Before The Next-Event…

Done…
Doing-again differently*…numerically…geometrically…*
Once more a looming endless cavity prison-cage cavern-entrance looking -out
Inevitably attempting to look -in:
-Heavy-Nucleon *lighter* and each distanced from each our own centre...
-Curious-stretched elliptic-orbital high-speed…
Around the rim-sides rhyming-verses:
-Hot-collided heated-points…
As if arguing and flaring then instantly-*cooling*:
-Cosmic Ray…

Cold-heated re-heating *movement absorbing as adsorbing necessary-fuel*
Sympathetic co-erced always willingly
With or without Degree of Choice against All-Other.
Of Free-Will to continue if not always ever.

As Us All *something more* than Us:
-Divine-*providence* and harmony...
-The-Captain remarks: Fateful! It was a cold and rainy night. The-Boson cries:
-I will tell you a story that you will believe!
-Every story ever told?
-Happened! I know *that* story!: It was a cold and rainy night...the Bosun cried: I will tell you a story that you will *not*-believe!
-Therefore when You tell that story again and again…
-You will be under my spell!
-With A Veil of Justice on-behalf of self and other but mainly-*Self...*
Translucent solid-reflecting or passing straight-through transmitting…
Transmitted sprung bouncing-along in Time as Nature.
Re-turning dynamic-processing

Diamontine-crystaline linear and *curving-ball*::

-*Monatomic Hydrogen gas...*

-Helium…*again…*

-Between Us and the *current* Cosmic Speed Limit...

-Absolutely-Zero...*cold*...nothing…again…*almost-precisely…*

-From Absolute-Zero *heavy* to this Absolute-Heavy…

-*Light…*

Hot fracturing-rupturing again:

-*Luminescence…*

-Spontaneously wavelength-frequency *determining…*

Each of Us ultra-rare atavistic particular-particulate

Crackling snappling-across:

-Cellular-bug *fleabag…*

-A regular electro-static cage of self-similar *irregularity...*

-Broken-fractured fine-scalar recursive-structure stochastic guessed at...

-Alignment…*pitted-pitched…*

Ahead and astern aiming aimed non-determined as a ship

heaving-to

Non-determining for the most-part:

-Random elemental fundament-tilting enough…

-Other heavier Z-boson capturing through the middle-knotted.

-Tau Taut-tightening...

-As the cookie-crumbles…chance.

Spoiling-negating passing-back through and around.

Each of Us lighter and heavier massing-massed energetic

Vacuum magneto-friction sticking *extreme* motion-machines:

-Kaptain Krump! Quantum Quark Krazy!

-Heavily-massing *baryonic*-Bosun…

-From-Rest *accelerated…*

Each of Us:

-Each Top-*heavy* top-Quark and bottom anti-Quark…

Heaving-*directional...*

-Impossible equality-assuming in-equality: The-Universe is not-*fair!*

-Fair-enough! Aiming closest or farthest distance…

-Mostly…*fair…*

-If that is what we expect! Intended-consequences…Otherwise? Why? If I am fair to You will most likely to be *fair* to Me?

-Intended see? Or *cheating…*

-But if i am and you are then?

-I am not and you then are? So, what to do?

-Done. Only two-ways…

-Or four or sixteen...

-Either anyway it is a Chance! Option? Choice?! Now! *and unless you are only in training and without psycho-pathic-resolve or otherwise unfairly rationing force over innocence you will be fair and so will I*:

-Win-Win.

-But...

-Unsure? 321...both-Families! Wed! Best of Both!!We know fairness when We: *experience* 4 by-*degree...*

-Of *fairness?*

-Except: when innocence meets force force meets innocence that is unfair to both sides. Extortion, exploitation cheating lying...knowingly-deceiving choosing-Self and Other which we do:

- All-The-Time:

-As if we never know after *The-Event* what it is we knew...

-Guilty All of Us!

-Of All.

#In that *memorific*-moment returned to differently:

-Nucleon-decaying entropic falling-apart...

Hanging-onto across-thirded balling full outer added integer-spin:

-Geo-desic geo-Metric:...
-Golden-ratio: 1.61803398875...
Coiled inward re-coiling outward spooling
Spooled re-turning angling:
-Each spinning-out half-spin electron negating electron-neutrino-fastest and smallest compressed-edges…
-Pine-seed seam shelling-stem branching fingers petaling…
flowering conic-container…
-As electronic *reverse*-laser...
-Neutral de-ionising Space-*neutrino* fired…
-Un-impaired constructor-photon...fired…
-Colliding absorbing-adsorbing…firing and fired-at as holding-apart…
-Firing-back probabilistic-resonating-*energy* mass-*illusion...*
-Chance-effecting causing-systemic un-predictable holistic consequence in The P*lasma-Field...*
-Free-proton electron ionespherical-paired universal reactive-*radical...*
-Paired compressive opposite close-enough for each paired photon to be absorbed...
-Charming and *strange...*
-Exotic colliding-annihilating light-scatter diffraction photo-*packaged...*
-Fundamental-particulate electrostatic-caged single-handedly directional…
-Nucleate electronic to neutrino anti-photon quantising quantised light-potential...
-Photon-gauge point fuse-field phase-*Symmetry...*
Across and amongst all-elements mass-energy gauge-symmetry:
-Wave-forming and with-*rapidity...*
Above and below and above and below *that*:
-Phonon-photon-fishes *phishing….*
-Longer-life…

-Only as memory-*intact* unbroken-connections stretching-out…

Overpowering hot-gloop brain-*freezing*:

-Quadratic chemical-compound every bit of information.

Letters and Space data punctuating like musical-notes photo-phono data-base computorial:

-Connective epithelial neural-tissue…

-Hyaluronic-acid potassium and sodium-*salt...*

-Ligament built and repair inflammation tearing granulating dust-bubble cell-migration…

-DNA txt long-code shortcode…

Binary –doubling quadrupling:

-A

-C

-T

-G

Wet-a-aware. Wearing hardware-goggles building:

-Played-out drama…in the Heat of Battle!

-Suffocation!

-Strangulation!

Each of Us encircling each and every other holding-apart:

-*I*sobaric triplet-charge neutral negative-positive raised again:

-Orbital-overlapping photon-electromagnetic carrying within each:

-*E*lementary proton positive negative electron-charge depleting...

-Proton-Positive Nucleon weakly-nucleating negative electron-leapt outer-ringing...

-Archer-crossbow string-line rapid-firing *pierced…*

-Taken-in absorbed *point pierced pressure slowed-potential phonon* repairing thudded booming cooling soaked-up bouncing-off at all angles refracting reflecting outer-tipped photic pixelated bosonic messaging-forces:

-With only *Absolute* Zero cold-certainty.
-Of *nothingness...*
-Within and *in-between...*

16.
Almost exacting-repetitive rectitude.
Following continuous containing easily-anticipating future-predictive
Imagining from memory chaotic-triangulate quarter-pocketing inside:
-By an arc of circumference passing through degree of arc-plane vertex-angle to the central trigonometric parallax:
-Tr-axial parabular...
-Zero point origin:.
-W X and Y Z axial...
-*Originating Trigonometric Complex* an inner world-forming turning involute universally and naturally limiting Cosmic-*evolute...*
Each of us stepped-forward dropped side-summing perimetorial comparator:
-Over-The-Parapet.
-Over The Barricades!
Fractious and fragile-formed We Each:
-Playing dice with The Universe.
And so...

Each of us a complete-gyroscopic of such *formless*-magnitude
As to be turning in and of ourselves.
With no *absolute*-upward or no downward and no inward and no outward nor sidereal facing:
-Universal-*encompassing...*
An encircling dark-pressure felt.
Thermo-dynamic skeletal triple-pressure point-phasing
Compounding cuboid converging parallel-piping prism-base
Stood or seated facing tubular:
-Triambic-isosceles:

-Hexahedral *golden*-dodecahedral rhombic:
-Probability-amplitude...
-Complex numerical-*propagation*...

Slipping-up or otherwise erroneously wandering straying
Distracting-forensics correcting with *nothing* solid-to-position
Within-and-in-between with only *sheer*-sensation…

Amongst-relative movement *darkly*-treacherous
Depthless cavernous and cold:
-*Absolute-Zero* cooling warming slowed moving sluggish…
-Now fractional-degrees warmer…warming…
-The-Universe.
Each of us each felt rebounded emplaced and replacing replaced
-Constantly-variably altering initial-condition...
Fearsome-expressing lost substitution-inserted
Maintaining-re-bounding:
-Inward axial-whirring whirling decay of death...
Defying stuttering-motion agonizing Anger!:
-Apex-predator wavelength-frequency…
-Ovoidal over-laid commensurate-atomic pressure-bar strutting…
-Delaying anticipating-memorific-associational
-Strung limbic-tailing distantly hooked Space-Spiders' Webs…

All-together at a distance each supporting Everything-Else.
Altering too. Feeding-fedback…
Each half-and-half-spin axis-rotating rotational energy-storing swapping
Conserving-producing immediately pre-vitally iterate
avoiding Each-Other as far as possible…
Alone attracting orbital-crossing:
-Electron-points *flashing* on-and-off force-fielding...

Roughly disc-Plate centring: Planetary-lunar roto-typical proton-positive pulsion...

-Hyper and *then*...

-Supervoval-Galactic compulsion-variating impulsion...

-*I*sotopical-diversifying spiralling-ionespherical...

-Atomic-*infinite*-potential *difference*-engine...

Felt-linkages and gears gauging sounded-out.

Sounding-out gurgling-gaseous surging-gradually and rapidly smoothly-out

Turning re-turning re-positioning re-defining characteristic:

-Re-iterating insubstantial-intermediary...

-Compulsion impulsing…

Heavily-hearted…

Each of us a new and renewing:

-The-Universe within each of us...

-Each *little*-universe...

-Multi-verse(s) within and without *constantly...*altering:

-The Universe: Each our own characteristic-personality *fuzzy* cult butt-crazy *buzzing…buzzed…*

Chilled-out frozen and warming swarming warp point-sustaining

Energetic wave-passed around and through each re-bounded:

-Cosmic-*red* ray-beaming Hyper-Noval!

-As disrupting-dice inward binding lightning *flashing* blinding…

Crackling-static:

-With mass charge-and-spin absorption...

-Reflection and disproportionate re-commencing...

-Incommensurate *commensurating*...

-The Universal-Centre:

Collapsing in-on-itself...

-Ourselves...

-How many?

-1 more than nothing energetic particles investing and divesting...

-Of *energy mass* only one-fundamental particulate of energy and mass...

-You!

-Me?!

-Why…not all energy? Why any matter at all?

-When?

-Why?

Angle-bracketed and vertical-horizontal bars holding-braced-across-ways:

-Because we decided it would be so and so far have been able to survive and show this to be fact...

-Particlefield-superposition?

-For now...

-For *then...*

Crashing colliding massless-producing:

-Paired-negation the conversion of kinetic-energy heat to light mass...

-Heat-destroying by entropy back to *energy...*

Cold:

-Returned: The-Universe...

-Made-*marked* Once-*again…*

-One moment more...

-When?

-All the difference.

-In the first fraction of a second of Time…Nature: *firing-off…*

Swarming:

-Universal-Protonic-*e*lectronic...

-Why?

Constantly switching-together-and-apart.

Each reduced and compounding-thirded-multiplying interaction

Each to each triple-quadrating *seeming*-infinitude...finitude...

-Why?
-Because there Is?
-...and We...decided it would be *so...*
-All of Us *for*-Ourselves...
-And Every*thing-else* that has so far *been...*
-To see? To hear? To touch?
-Being....and...able...to *survive* to show this to be *fact...*
-All of Us? Everything? Why is not All-Energy? Why Matter at All?

Power and Light! Universal-Energy every moment colliding mass-*producing destroying...*

-Re-producing...
-Nothingness?
-Life! Violent forward-Force...Forcing...
-With minimalist-serenity...
-Nothing at All?
-False. The conversion of kinetic heat energy...
-To-Mass...*forcing...*
-Change. Changing-colour charm and...exciting...

Once again changing-direction and velocity:

-Exotic. Speeding-up slowing-down accelerating and de-accelerating...
-Strange. Frequency-wavelength *effervescence......*
-Quintessence! QuantumQuarkKrazy!

Our Own Techno-Machination-made and *Driven...Chasing...*
-Hidden from *unknown*-rest...
Accelerating on the re-bound
Rogue-rouge:
-Live Evil!!
-Let Evil Live!
In-reverse black and white and blue-green
Or neither:
-*Freak*-wave!
-Quantum Quark Plasma-Zoo!
-All-Colour! Primary to secondary and so on...
Space-stretching frequencies validating variating-*vitiating...*

Immediately heated-cooling -gaseous crystalline-solidifying-watery *almost*...
Each of Us:
-How many?
-1...more than nothing...
-*To at least*...762559748498...
-And *counting*...
-The *observable* Universe...
-Measurable light at 5% the rest as darkness...
Dropped-in *top*-heavy Quark-breaking breathing shattering-apart made.
Inside peaking and bottoming-out
Boating-*heaving*-directional:
-*Strongly*-Nucleating nucleon-decaying entropic-*neutrino*...
Falling-out. With-Everything...from as into-*Nothingness:*
-Hanging-on to the geodesic=geometric...
-Each potentially infinite uniquely linear-circular...
-Finite Golden-Ratio rounding up and down...
Rationing realising releasing coiling-outward re-coiling doubly-spoiling re-
spooling line-netted:
-Re-turning thetical and anti-thetical:...
-Stress electro-magnetic plasma-field:...
-When?
-Now.
-In *that* moment of Time?
-Then? Naturally...
-Deceptively...
-Fallaciously...
-Knowingly?
-More or less...
-Us?
-Naturally...
-Now?
-Then...
-All the-Time?
-And now time's...

-For how long?
-Immortal? Eternal?
-To-then...and now:
-Why?

In the first fraction of a second of Time:*Naturally:...*
-Firing-off...
-Fired-off.
Angle-bracketing vertical and horizontal and diagonal-bars:
-Particle-field in superposition warping...
-Why?
-Well because we-could...
-And because we needed-to have done...
-More or less...to-get to-Here? Now?
-Correct.
-Then?
-Correct-too.
-And because we *could* and because we *wanted-to*...
-Too...
-To?

Enlightening discarding discharging more than the cubic-equational sum-rule:
-Ever more than the energy-parts *needed*-to fuse form larger-atoms: di-Hydrogen deuterium...tri-Hydrogenic...
-Tritium...
-Self and Other: molecules and other *self-serving* cells...

Each-uniquely ion-spherical
Trait-stating massively *swarming*...re-warming ruffled composed:
-Dynamic thermonuclear *transparency:...*
Each sprung-circuital warm swarming-fusion reaction-split
Self-replicating altering-automata
Mechanical-computorial re-registering-state:
-As disrupting-dice inward binding lightning *flashing* blinding...

-Non-local unpredictable complex! complex!! Complete-template...

Replication-copy of Self and as sticky-glue taped-together combining...

-Proteinous-looping programmatic...

-Unit-listing constructed-constructor spiral-arms and legs...

-Any error could be disastrous...

-Selecting-out...

-With few mistakes...

Heaving in the parabolic-panoply finitely-infinite possibility:

-Quark-*Quintessence...*

Interfering patterns-of-reality:

-Starting staring starring electronic-*positronic* switching-circuit:

-Each phonon-winded and photon-particulate potentiality
collisioning...

-Absolutely with each other and all else then...

-It happened...

By then had already *happened...*

2. The Universal Axial:

17.

-Now! As we now know it was not a pressure from outside for in the apparent emptiness of *nothingness* there was *none...*it was *the* sheer weight of everything inside distinctly distantly and singly spaced together, and apart...

-That darkly and increasingly heatedly weighed down and gradually squeezing crushingly-*weighted*: *The Universal heated Thermal Tipping Point...*

-The Universe!

-Let there be l*ight!*

-Everywhere!
As all around
Overtipping minimal-*maximal*:
-Universal-*heat* massively-cooling and re-heating...
Collapsing-crystalising centring and dragging toward the dark:
-Universal-Hub.
-The Nub of the matter...

Centring *individual freely-naked* with each nucleon-fired
With each other alternating filing-fielded each:
-Black-Star dark-energy and matter...
-White-Star *light* energy and matter...
Guttering denting dense-channelling furrowing funnel-tunnelling
Openly- *corrugating* coruscating crossed-wave-lapping.
Folding and circulating fluid glittering forming stellar:
-Quantum Quark-Oceanic...
Shredding and slotting-into.
One volatile state continuously into-another:
-Wave-tube *wormhole* transforming...*sliding...*
-Electronic-*information*-energy cloud-smearing...
-Each wave-stellar wearing core convection-rings...
-Spiked-corona planetary inner-core outer-core…
-Mantle and crust: inner and outer-rings layer-upon-layer…
Thrown-out into orbit-*coalescing:*
-All the same-*aged…ageing…*
-As from The-Start: The Beginning…
-As Stars and Galaxies and Planets and Moons…
-Asteroids and comet landed and *whizzed* by to the inner and outer Asteroid-Belts *figuring*-relationships…

Wearing weaved solid-state-particulate egg shellaced
Magnetic as iron filings *leaded* weighted-down *molten*-silver to golden-dust rocky-core

With enough gravity to make a rounding watery icy-mantle
thin-atmosphere:
-EarthCentre!
-Planet-dust!
-Lunar-rising...colliding...
-Disintegrating *flash*-lighting explosive-burning lit from within full-out surface-*backwashing*...
-Catching crossing a curve curling-wave!
-So-far! *e*mergent...
In the lightening dark-lashing mists burning-off:
-Coronal-headed lowered-louring lancing-pointed chased...
-Conduit-Street...
With the flimsiest of fuzzy-glimpses *ghostly*-glances:
-*D*isappearing quanta: SuperFluid...
Washing quantum-bubbling foaming twisting and turning
Each bridged-snapping and spinning-StarCar conjoined:
-SkyWheeling!

Switchback looping-the-loop twisting multiple suspended-chain swing-boated Oceanic-seas.
Each of us spin-wheeling vibration-rattling coasting warp-and-bent
As from all sides cupped-suctioning at any moment positive-curvature
Negative double turning inward as out:
-All-together...
-Universal-Engine-Block *frameworking...*
-Drilled-into.

Ghostly-glimpsing only the shadowy-outline cadaverous
Dark-ceilinged open-skied mountainous-conjectural
Holding-cavernous arrow-dynamic-directional
Free-wheeling flip-flopping-over:
-Tiptop-Crazy!

Flat wave-riding darkly smoothed-out:
    -Fractal-almost...*almost*-repeating turning...
Cellular-bubbles of seeming-air
Gasped *gelatinous-liquid* fleshed together toughed outer more solid-skin Ultra-violet melanoma bone-forming over
Connecting-tissue within Each-of-Us *becoming...*
Each lumpy image-*flashed* blazing-*dizzyingly*
Giddyingly spinning-out organic flywheel launched:
    -Inner-*vacuum* vortex-backdrafting *helicoidal-tumbling...*
Centring:
    -Helter-Skelter Madness!

*Darkness...*dissolving near-vacuum bubbling-along.
Cloudy evaporating white-cap energy-edged de-compressant:
    -Hyperbolic-conical compressent *i*nward-falling freely outward...
Flung inter-locking *ambient*-spectral:
    -Big Wheeling now!
As water-wheeling core cogged working clotted clogged loosed uncogged: kinetic rousing carousel carousing runaway unstoppable...
    -Roller-Coastering!
    -Sky-coasting! StarCar!
Jump-started from *nothingness...*
    -Cosmic-FunFair! Fairgroundshowground ride…

Into The-Void: *emptiness:* Theme-Park *distending…*
Whole-Motoring whirligig.
Hot-gas-ballooning rig-risen loaded…
Carrying-caged drum-spinning:
    -RadioTronic...
    -Rotary-Gravitron:
    -*Light*-Synchrotron...
Centrifugal para- as dia-magnetic
Ringed-poles colluding:
    -Uncertain we crawl to Death!

-The *definitive* reigning-in...
-A Lottery!
-Two-step syncopated three-step four-step at the helm.
-Each *illusional* Gambling-*wager*...
Bluffing cheating to get-on or get-out of there
And in equal-fairness colliding-and-colluding-avoidance mostly:
-On All-or- *Nothing*...

18.
Trusting to everything proportionately calculating:
-The Measurers!
-Rulers! Now!
-Oh Reason not the need!
-The Decisive-Act! Action! Actions...
-Each-Act *always*...
-One-step ahead of barely conscious *thought*...
-Complex-timing-motion throughout...
-Naturally in Space!
Clicking asymmetric in-form and function moving.
Rightfully as Responsible de-claiming
Instinctually morally-accountable now
To Self at least and Other.

Felt-*only* or unfelt unknown
Ever-consciously in-retrospect:
-After-*The*-Act.
-Extreme-Ratio Arena:...
-Receiving-*fairly* not necessarily-giving proportionately *fairly*...
-Given at some point continued numerically up-and-down...
-In-and-out and through-and-forward and-back almost...never exactly perfectly:
-The-Same again...

-As almost-*equally* as necessary…constantly-changing…altering-successfully…
-Socially. Carefully Risk! *Averse…*
-Of Ourselves and Other…
-Taken for-granted…
-As granted…yet *essential*-Absolute and gone *varieties'* of complication passionate not compassionate and *irregular…*
-Self-constructing...
-Universal social-construct...
-Instinctual re-action and *red*-action *mourning* the blue…
-Personal-*emotion* coming-down coming-out and through and into:
-Being:
Tearing at the imbalance to balance
Un-balancing:
-The-Universe!
-Life!
-To The End of The Universe!
Critically-pointed acceptably tipped-into:
-Kaos!
Three-bodied charioteer in all-directions and one towards and away from:
-The-Start! You!

Reasoning only *after* The-Event even as if *intelligently*-designing with *all* potential-probabilities based-on prior:
-Knowledge of Each…
-As Knowledge of True-*belief*. Information.
-Of *each* experiential experience-fading into almost vaguely remembered-*belief* True or not…
As each moment Super-*ceding…hyper…*Each-Other…
So quickly as to seem slowly smoothed-out.
Filleted and filling flinging in the gaps filed-outward:
-Free Proton electron ionespherical…

-Compressive attractive electrostatic-force colliding exploding cascading
Stumbling-inward and by outer-degree:
-Toward and from what we now know positively to be:...
-The Centre of The-Universe…
-Gone.

Each-interlinked force-field with everything-else and throughout:
-Critically over-weighted…
-Massively Universal-Hyper-Star...
-Galaxies…and Universe…

Formed-imploding an emptying empty hollow-centre:
-Folding hyperbolic-twisting unspiralling helical-pointed ending exploded open-closed…
-Universal Proto-star! Hyper-stellar then…
-*Super-Noval...*
-Energetic-*shockwave...*
-Universal-Forcefield...
-The Edge of The-Universe...
-Our Universe! Universes…
Each of U *spiraling* almost *out-of-Control*
Unable any longer to hold it all together: *i*n newness...
Information-alighting lightning-lighted
Clashing-reflective absorbing Thunder! Reflecting
Reflected adsorbing-absorbant pressing-out again.

19.
And onwards: *testing*-invention discovery and design
Re-discovery experimental-*togetherness*
*T*ogether each-momentary:
-Finite possibility potentiality and probability...
-Of What is and what *may* be ingenious atomic engineering...

-Super-*engineering...*
-Hyper-manufacture Supernoval...
-Planetary-lunar: EarthCentre!

Postillion *Intelligent*-Design yet *seeming* more than *that:*
-The Technology of Nature...
-In-Time...

M*emorific*-metallic experiential-machinery:
-Experimental-memorific...
-Mechanical-energetic particle-*investing...*
-Divesting of energy-mass...
-Exploited-exploiting...*deceiving...*
-Extracting-extorting...
-Force-and-force merging matter emerging...
-Innocence emerging as what we are...
-Self-interest...
-And what will be: Other.
-That which holds it-all-together and blows it All apart.
-Gravitational dust-grains...
-Light-*potential...*
In-*newness* inside-out alternating breathing-pieces in-and-out:
-Constant squared-circling valiant-variant...
Falling-off at less than the inverse square of the distance created:
-Pulsing *Pulsar...*
*E*xpansive and contractive contraption breathing-parts:
-Feeding differential-motion heated and distanced...
-As red rare raw *cooking cooked breathed and drunk...fed...*
Through the yet still dark only imaginable gloom darkholed:
Quasar-ring *radioing...*
To the *virtually* silent-outside swallowing gas and winding into the *gone* centre spiraling magnetic-field lines-regulated re-*regulating...*

Accelerating electron-positron captured and released in almost-perfect shifting-homogeneity:
-In all directions equaling out-again...
-Never the same again...
*B*umping-along irregular-edged etched-bordering
And only now inexorably being-pressed together
Colliding crushed and crunched together:
-Your own-owned Atomic-*clouding...*
-Between *nothing* but the fine-fundamentals...
-Travelling between *sparticle*-energy carried-carrying...
-*I*ndividual and unstoppable light-show immateriel:
-Synaesthetic-synaestheatre...
With a hail of fluorescent hard and soft W X and Y-ray
Photon-bombardment bending light and dark-radiation
Disappearing…
Winded dawn-drawn drawing-out so as to disappear into:
-T*he Black-Star of The Universe...*
-The White-Stars of Us!
Without mass or spin parts with mass and spin…
Nucleating collided with:
-*Each* your own gravitating attracting and repulsing...
-Starry Galactic…
-*The*-Universe!
-Our Universe!

20.
Moved-around electronic-circuits bridged-and-gating resisting-potentiometer:
-Other but still more each to ourselves complex-membrane mirroring skyward…
-EarthCentres Stellar re-bounded…
-Of every other Colour…
-Each of us a *sudden*-revelation each-to-each and all together
Reflecting reflected incidental accidental rays
Angled equally co-planar drives

Desire and repulsions finding-unity in the image
Of other each our own specular image reflected-off Other:

-A guide beyond the imaginary…

-*Signifier* finding *the symbolic-symbiotic…*

-*The real* fizziparous bloviating obviating...

-*Imaginary*-games of smoking-guns and mirroring duplicitous duplication…

Doubling-projection
Slightly-altering or of reverse-identification:

-Each and *every*-side radiating photonic-precipitous...

-Real authentic inside-integral to the known-exterior unknown...

-*Conflagration* condensate-*rivering*:

-Oceanic colour-wavelength frequency positively-charged parton photonic-primary packaged-points of *light...*

-Giving and gifting some to another re-gifted...

Reflecting-off of *densely*-clouding:

-Condensate outer-splashing coronal-spin foaming...

Evaporate-bubbling falling dark-noisy *crackling*:

-*Cosmic-Microwaves…*

-Back ground-*radiation...*

-Us. We. Ourselves…

21.

Abstract algebraic-geometrical hollow commutative-ring ovular:

-Prime-*ideal* set morphism continuous leaf sheaf schematic-spin-*weaving inside...*

-Categorising functor-mapping universal-factoring intensity:

-Ovariation no-more than 99.999 variation...

-Mass unseen dark-matter...dark energy...

-Right now?

-Right-now...

Crashing into another *almost*-accidentally...

-Egg-shaped three-dimensional...

-Arrow-pointed fourth…

-All paths and points *relative* to each other...

-On-purpose or purposely re-directed...self-directing...

-Net-force required to accelerate towards or away…

-Move an electron out of one-atom to freedom or entrapment by another...

-Other-directing…willing…together…

-From *neutral*-Hydrogen singly uniquely charged between iconic ionic-parameters...

-Lakes and seas...

-Mountain-streams...

-Planet…

-A mass of one kilogram one click per solar-second per-second...per-second +/-...

-Electric-charge potential-difference...

-Producing one-Watt of Power per 0ne-watt second:...

-Cosmic speed of light in a vacuum 1/300m stepped run click seconds...thirds…quarters…

-Mass per one litre water click-cubed duration of none-period orbitals incomplete continuing wave-oscillation...

-Dimensionless cancelling-out...

-Radian: *The Radiation of the Transitions...*

Between two-hyperfine-levels of the ground seemingly static-planars meeting:

-Engine-plates effecting affecting *vacuum*-fluctuation...

-Net-force uncharged virtual-particle photon neutrino charged zero-point energy:...

-Sum-attraction and repulsive dipolar-forces...

-Co-valent-bonds electrostatic ion-interaction:...

-Structural-surface polymer-condensed *e*-vaporating quantising-field...

-Non-polar to polar electro-negativity-necessarily *asymmetric-moving…*

-Structure-conducting conductive-tunneling current stellar wind and wave...

-*Electrons* passing across together banded and taken taking-in...

-A point-per-second *luminous*-efficiency *function...*

-Intensity mono-chromatic-radiation frequency-radiant...

-Iintensity solid-ray beam in 3D space: hexagonal line the smallest distance between two-objects *subjectively...* Solid-angular span positional radian dimension-full:

-As -minus square-cubed +plus…squared cubed-all=1 radiant-exigent exitent in-all-directions irradiant...

-Black-Hole to White *Light*-Wall…

All around shattering yet further beyond burst-puncturing penetrating again:

-The Universe!

-Ourselves! *radiating-sources...*amount of substance matter=energy equivalent: molecular-mass paired a dozen fundamental carbon particle +plused /and -minus hyper-cubed...

Around the *frosty*-edges de-frosting felt raining flaking-searing scattering-clicking *flashing* points...

-Radiant-*radiating...*

Fountaining-firework showering detracting-sprinklings prickling point-like flowing:

-Condensing *inconceivably* massively-heated...

-Evaporate imploded and exploding-again: shattering-apart.

22.

Blasting:

-*Neutrino* pole-to-pole:

-*The Universal Axial...*

Flowing flowering-momentarily upward as well as downward
Sliding sloped loping in all directions-away...
Only one toward at a time each tracking each other

Colliding-*photon* firing rivering…
Oceanic *c*omparative ionic-lopped
Spiked-and-rounding bowl *p*eaking and *troughing trim*:
-Each-electron charging charge vertical and horizontal spinning-spun...
-Unavoidable pushing and pulling…
-H*yper-dense neutron star-collapse* drowning:
-Zero-point holding collision repulsing and attracting...
Inside-tunnelling funnelling fountaining.

Re-assigning matters through venal-flow *pulsing shock*-waves
Each and every one of us. In The Mix. One Over The-Limit.
Which One we may *never* know...

That One co-ordinate each converting each and every Other.
Clashing crashing colliding runaway reactive taking-in
Knocking-out the greater vector inverse-field proportional:
-To the *inverse* cubed-distance...
-The space reducing in-between directly-proportional...
-To the product of the magnitudes of each and inversely proportional to the square cubed the total distance between Each of Us:
-Ellipsoid-equatorial...
*D*e-stable blown apart cataclysmic-catalysing...intense split spheroidal-magnetic-field spin-fluxing:
-High-energy particles passing *straight-through…*

*E*xpelling and sponging primordial-swirling:
-Photonic-*light* hot swishing swatching switching-energy switch:
-On/Off...binary…
-Chiral-and-sinstral reactive-forces...
Within. Almost opal-*op*aque not pitch-black any-longer.

But pitched a shimmering *light-Stellation...*
Imploding powder-black blasted-back from core-to-surface

Surface-to-core *rippled* ahead and backwards:
-Once more close-to…
-The Universal Cosmic speed-*limit...*
-Which-is?
-Any speed except at the centre limiting-*lighted...*
-Photonic dark-centring electro-magnetic *gravitational-* ionespherical...
-Poly-Mono-Atomic...
Switching neutron-paired-proton core-collapsing:
-Supernoval...now...
-Hypernoval!

Compound Stellar-Galactic splashing super-conducting super-fluid
Swim-wave absorbing adsorbing and reflecting bouncing-off…
Bubbling-foaming centrifuging-capturing centripetal briefly:
-Swollen bodied-breathing:...
-Energy-bellied pollen-feeding honey-bee…
-Other...
Articulating expressing in each winded breath and each exhaustive-fart.
Carried-off on the wind or off another cellular
Re-storing circuital swing-seated
Quadrilateral limbic-held penta-tonic
Hand-and-foot gripped; grasping-grappling
Uncertainty at the controls
Full-body-Game Control-System (GCS):
-GPS (Global Positioning System)…
-The moving and positioning of each and every point-piece at any stage...
-What is allowed and if at All-possible...
-If at all *likely...*
-Without: The *Rules-of-The-Game...*
-Rules?
-Of-The-Game?

-Made-up as we go along. Each on our way...

-In)ur Own Way to: The *set aggregate of all relational properties...*

-The *rules* constantly functioning...

-Always doing what we want...

-For or Against...

-Constantly-changing continuously-stop-starting...

-As to how we each arrived each of us in this personal and conditional-state. Each of Us *exclusively* occupying combined-composite:

-Spin-wheeling around in all and each and every direction... looping whooping:

-*Quantum-Complex* selective-testing: on-off switching tick-ticking click-clicking clock clocking stumbling tumbling up and down.

Each of us *individualistic* distance-timing-velocity

-Odometric-potentiometer...

Jolting-turns flicking *flickering* changes. Of Each. And Each and every Others' assumed felt relative-speed

Singular-direction precocious-building...

Precious and with prowess combative at each momentary-degeneration

As *a* re-generation...

As a continuous containing presence facing continual annihilation resurrection.

Within this and with the outside understanding of more than all this

A feeling of more than all this...

Dark energy and Mass-*uncertainty*:

-Universal-*stabilising* single-simple multiple-switching...

Discarding-*discharging:*

-More than the equational-sums of all the continuously existent-parts...

-Perpetual-struggle smuggling continuing...

-Life-*Principle...*

-Scientific-protocol…

-This *Personal-space*-within and part-of-IT-all…Public.

-Out-there...tending toward life as life tends toward death. Once existing always to have existed...

-Everything-more than the sum of part or parts...

-The feeling of something more something more...than...

-Other?

-Everything-else! The Universe! Experience of this consciousness and individual-existence...

-The *integral*-Set of Everything!

-Universal!

-The *feeling* of *something* more...

-Hyper-Space?

-Cyber-Space?

-Super Tri-Galaxy!? This is IT! Isn't it? Grand-Design? Grand-Designer?

-Designers! All of Us! That have ever been and will be...

-Who? What?

-Yours. And yours' alone...

-What? Mine? All of it?

-Yes; and Ours...with Ourselves and...

-One more than the-total-set?

-You! Me? You...and Me? Don't You see? Listen: All of Us...and...

-Universal! Galactic! Star!

-Planet!

-Moon!

-Terra-*forming*!

-EarthCentres!!

3. Universal Axial-Stellation

23.

Over The Cosmic-Horizon.

In the moment-selective thought-provoking imaginative-memorific
Decision-making: a certain re-stabilising
Habituating-memorific story-telling…

Of *meaningful*-activity leaning this-way-and-that wobbling tilted-together:
-Hell's Pit!
-Hovel!
-Dump!
-Slum!
-Shot-Gun Shack! Pig-Pen! Pigsty! Horses-*hooved…*
-Liar-Lair!
-I only ever speak Truth! This is A Lie!
-Burrowing-run warren-sett…
-On the rack of this wracked world stretched!!
-Heaven's Vault!

Randomly-*seeming* whirling-orbiting cupping-spincars
Variable combined in- motion unexpectedly swung-outwards again…
Snapping rotating and disconnected and filling:
-The-Whole of the only Personal Perpetual-Space…
Rippling-energetic embedded embodying full-felt effect conspiratorial collaborative *at the same time: silver*-grey fogged *plasma*-clouding golden-*glittering…* gaping dark-energetic-matter-felt as sono-luminescent *booming* blazing livid-*latency*.

In rising-mists shifting soft-white silvering altering colouring-chromatic black to white and in shades of *grey…*thermal-light between long-wavelength legs infrared electromagnetic:
-X-ray short wave high-energy ultra-violet...
Between and amongst:
-The-Universal prismatic-*chromospherical...*
Lighted-the-visible-spectrum *shimmering* red-hot streaked-Gamma!

-Reflected shining dark-brown orange and pink-patched...

-Luminal red green and blue *photon:* Y-ray Gamma-decaying...

-Spectrum-wavelength probability-function collapsing-configuration...

Conflagration! *Integral* each of us made burned forged with the full force of *imperative* impenetrable-initiating a *critical-Course-of-Events...*

Each of us critically over-tipping thermal-point *quivering...*
Bubble-popping arrow imploding:

-High-pressure heat-fusing bursting with relative nuclear-furnace:

-Protean-Protium...

-Proton combining Hydrogen-*gas...*

-Ionespherical di-protonic *liquid* Dia-Hydrogen compression...

-Simultaneously releasing positron and e-neutrino:...

-Annihilating light-energetic laser-*photon* exit and egress emitted...

-Entitled exact numerical-neutron anti-matter proton material conjoining:

-Bi-protonic: Deuterium-fused cooling-binding energy...

linking unstable-gaseous liquid now:

-Para-hydrogenic: TriHydrogen...

-Tri-genic double-neutron nucleon-Tritium...

Fusing-binding energy-linking un-stable gaseous compressing liquid-*slush:*

-Tri-Hydrogen...

-Hyper-Hydrogenic...

-Titular-Tritium...

Extremely-luminous instantly pledged at the still *frozen* edges:

-Pairing dia-magnetic electron *honorific-*
Helium...
super-fluid flowing super-engineering octillion cotillion
postillion.
Extremely-luminous at the extreme instantly *unfozen:*
-Paired-proton electron...
-Helium-*superfluid...*
Flowed Oceanic-rivering slow-moving it seemed against all else the same
So
Stilled while hurting hurtling-along: **somethingness** overwhelming neutral-*nothingness...*

24.
Internal-organic external-energetic purple-glowing:
-Electrical-field superconducting without outer friction or *resistance...*
-Orange-crust core supersolid crystalline mineral-*metallic...*
Dust-clouds brushing main-sequencing seeking stabilising-paired:
-Full outer-electron shellhole dipped ringing sub-shell lords a-leaping:...
-Outer orbiting dis-associative protonic-electronic second-electron-shell...
-As freely grabbing de-stabilising Lithium branching conducting conduit...
-Alkali-metallic trace impetuously silver-sparkling...
-Combining double-pair ringing quadrupedal...
-Bivalent Beryllium alkaline hot solid earth-metal puree...
-Clear blue-grey halogenic elastic tri-ring black brown powder-grainy
-electron trivalent-quadrilateral verticular and pentangular:
-Boron ...

paused; metalloid gaseous-liquid waste-exhausting:
-Cold-fusing held-up stable-waiting thinning-rapidly expansive-delaying inorganic-decay:...
-Disordering electro-static repulsion-hovering holding-apart moved further-apart...
-Disordering-decaying *eventually* collapsing heat-combining compressed massively re-heated kinetic energy-tunnelling wave-function *flashed*!
-Helium…
-Runaway!
-Runway! Taking-off! Taken!

Lit fast-tripling emergent-meeting *meting-points melting…again…*
Each of us belted and buckled in place hard-hexing organic-and-incomplete:
-Carbon-fibre cycling...
Re-cycling resonating ordering-quadrilateral tetravalent Co-valence-chaining:
-Nano-nucleo hyper-Stellar Synthesis:...
-Another hydrogen-atom…
-More!
Adding grabbing outer five-shell clouding:
-Nitrogen sedimentary rocksalt silt-dusting planetary-stellar nucleon-thermionic electron *clouded...*
-Insoluble watery refractive: acid-watery light brown-alkaline solvent flowing-transiting:
-The Water-*Bringer...*
-The Acid-Giver: six-shell…
-Hydrogen-Oxygen watery forming foaming gas-liquid-solid *homogenous:...*
-Carbolic-watery solution fixing-liquid filtration…
-Hydrocarbon...
hissing and Sizzling sintring carbide-fired powdery-multivalent:
-Nitrogen-group grouped...

-Oxygen paired-binding catalysing burning-across dulled-blue carbon di-oxide...

-Tri-atomic compounding...

-Energetic co-catenating long-chain door-gateway...close-up:

-Closing-down...inwards and outwards...

-Ammoniac-alkaloid: nitric amino-acid colourless crystalline sweet-tasting...*glistening* metabolic potentiate stellar-spectral synesthesia synthetic-tongue ears and eyes looking-out listening-out for sweetness-souring wavelength extreme-frequency attractive pink-noise deeply-resounding red-blue note shifting noisy purple-red to *grey* seen:

-Helium stellar-gas and *ash...*

Crackling *super-imposing* chemical-splitting hydrogen-bonding feeding-fused lighter to heavier combining:

-Isotopical side-arms scooping up legs-kicking...

-Odourous tasting electric-burning smoke-screen shielding-clouding:

-Souring-acid added toxic-*poisonous* darkly-adaptive grey-blue...

-Absorbing-gaseous molten metallic glowing halogenic flowing:

-Fluorine.

Stable full outer shell-valence highly-reactive:

-Octagonal-orange-glowing:

-Inert-mon-atomic lightning bolted noble...

-*Neon* lit-up.

25.

Orange-paling solid-*salting*:

-Caustic-soda hydroxide...

Frothing fuzzy fluffy tufted fibrous balling misting whispy noval borders cusping scrimming:

-Sodium-soapy oily metallic-conducting registering felt-moving building-up electric-charge...

-Felt-movement mid-point lit...

-Milky-Mechanical frontward smell and taste arising ashen-appetising...

-Or disgusting dis-guising cerebral-cortex core-*sensational...*

-Cortextual-contextual recording-over-dub... reflecting-comparison messaging dissipating-difference:

-Warning!

Flaming soft-*silvering:*

-Magnesium-*flaring*!

-Hot-traced Earth-Centre metal-filigree...

-Silver-white astral-Aluminous glittering enamel quatrine-quartz metal solidifying:

-Silica burned-onto and together circuital...

-Tetra-valent squarish-shell white yellow red violet blue black...

-Non-metal: Phosphoric snowball phosphate-fertiliser...

-Sulphate-first arsenic-poisoning Antinomy: volcanising-rock:

-Tellurium...

-Odorous Iodine stinging: Spectral-Xenon.

-Caesium: *toxic* and *deadly*-poisonous strengthening-spinel ever-circling skeletal-familial:...

-Chemi-*luminescent* phosphorescence-phospholid...

-Hydrogenic-headed hydrophobic-carbohydrate strand lining-up head and tail flowing positive-negative *felt...*

-Molecule-membrane channelling-liquid metal-mineral paths...and doorways...

Opening and closing:

-Protein-lipid polymorphic bilayering-phosphylation energizing valves...

-Activating or de-activating as yet inert bright yellow anaerobic:

-Sulphur...

Scalding brimstone-volcanic erupting-golden-uric pale green:

-Chlorinous...

-Salting-bleached sand-beached holding resistant…
-Non-corroding *Noble*-Argon:...

Inert and unreactive aery-atmospheric cloud-filled second and third...eight-ring planetary orbital outer-octet valence-stability:

-With Hydrogen Carbon Oxygen and Nitrogen...
-With Sodium and Magnesium metals...
-Internally and externally extremely complete-rings...
-Ionespherical and covalence-linking corralling-core...
-Solo and dueting...*waltzing:* Hydrogen and Helium mainly…

Floating as Lost-in-Space

-Switching outer-valence ring inner binary-base...outer-planetary

electron *dropped*-inside:

-Collapsing de-stabilising subshell innertrack-centring fields filed around...
-OK! You got IT! *Free*-electron capturing!
-Captured!!

Vacant empty zero-gap filled added-to.
In-between accreting planetary and lunar equatorial dust-complex clouding:

-Acidic-fracturing corrosive-liquefying piquancy...

As if *goading*-reaction:

-Igniting *hydrogen*-salt watery-slimy…
-Only *slightly*-soluble...breaking! The *Surface*-tension …
-Soapy-lipid pungent unstable-Potassium-ash flossing...
-Foaming-over crystalline grit-grain:...
-Hydroxide peroxide-potash sprayed limbic-skeletal...

Dully-silver string-strung inside-undissolving:

-Winding-Calcium...

Lattice-frame binding-bonar skeletal metal-mineralising processing fourth-shell:

-Stable-centring Helium vortex-pair filled...

An additional electron-captured another-forced onto the inside track:

-Elaborate Ekaboron outer electron-pairing:
-Helium inner-shell screeching...screaming: not-inert!
holding-collapsing crashed-into *de*-stabilising...
Dropped-inside again outside transition *metallic*-threaded:
-Toughing Titanium...
Violent fiery metallic colourful:
-Vanadium...
Single outer-ring planar-planetary plate polished sheer podiametric:
-Chromium-lighted quashing hydrocarbon *flashed!* outer-hydrogen ring…
Transiting-*silver*:
-Manganese-pressing electro-mechanical reverse-flowing...
-Induction-charging...
-Weakly-pumping against *nothingness* and each other...
Around-and-around:
-Ductile metal-added stabilising outside-tracking purple-helium aqueous-solution...
-Epicene eponymous aqueous-solution:
-Magneto-spectral...
Evenly unevenly in repeat-sequence overtipping:
-Cobalt-blue *piezo*-electronic magnetic-charging spectral continuous almost symmetrical...
-Differentiating-manifold commutative-*automorphic...*
Switched-lanes filled-gaps gradually-stepped stabilising-reducing and adding up and down continuously-chosen in and out:
-Ionising…
Holding and sticking transition metal molecular conjoining:
-Sky-Wheeling!...switchback-looping the-fractal almost-repeating...
-Each of us become...becoming:
-The Universe!
As our own:
-Solar-Systemic...

-This One!
-Galaxys' eighteen complex completing magnetic-ligament chained:
-This One!

The most *beautiful ambient* honey-comb gapping:
-Reduced nine electron orbital positive root cubed lattice…
Captivating sublimity axial tri-line parsing:
-Prime-binary…
-Polar-irregular unevenly-tracking...
Tracing:
-Each a twenty-seven electron-point stacking…
-Hyper-Cuboid...
-Universal-Holograms…
Outer-shell holding solid-setting:
-Nickel...
To mutual-mutant repulsion of so many positively charged-particles:
-Too strongly *overwhelming…*
-Ferrocene: iron-pentacarbonyl ion-decaying outer-rings...
Reverting astro-metal super-chilled crystal-parallel chain-spinning electron bar-magnets:...
-E8...248-*dimensional...*

Centring super-symmetric matrical peak grouped: *iron*-bloodied
iron-willed iron-blocking:
-Universal pressure-voiding heavier tougher and less likely to be shattered apart again…

Holding-together as breathing-in quietly absorbing adsorbing more:
-Atomic-*energy…*

Than can be released; crushing suffocating crashing technological-collapsing:

-Atomic sub-shell valence-rings matching stable…

-Outer-shell at variating hyper-surface unmatching exchanging…swapping…seen-circulating dulled-graphite glistening glittering point-diamontine:

-Universal-Hub...*nub...*

-Cosmic-Core! Your-*own* temperature-gauge-0+ metering...

-Measuring! Ruling *electron* in or out!

Wave-lapping folding and circulating through *molten* hot-metalline.

Crystalline-starlight scattering scattered prolapsing spiked-tipped:

-Fractal-framed movement reducing outer coronal bubbling-edges

Speared eclipsing-eclipsed:

-Blackened Carbon Iron-*steel* stolen ball-bearing circling squared-triangulate...

-Cornered rapidly-reducing:

-To Helium *de-stabilizing...now…*

-Ionising *radio*-carbon decaying digital...strung-stringing together...dis-accreting neutrino-dust iso-spinning helical-combining:...memory…*data-storage…*

-Each of Us! Atomic-*molecular* stabilising-spinning parallel-chain-carbon steel-plated nickel…aluminium-*silver*…Gold!

-Outer and inner *solidifying* super-conductive:

-In a *vacuum* fusing ferrite-*spinel...*

Starry-sky crystalline iron-grey clouded sheet heavily draughted over each as if never been never-been never-having *being…*

Each One of Us star-speckling otherwise *vanishing*-points…

Utterly completely completed blanket-covering

Breath-holding held darkened-*pitched into*
Back-tracking naturally to an *almost* timeless-*nothingness:*
-Anti-electron positron…
-Anti-Everything!

26.
-The Centre of The Universe…
-The-Edge…
-Absence-of-*Light now…*
No-sound no apparent-movement.

In absence of further processional-movement.
In no time to be switched-off.
To disappear completely
Into a dense deep centred silent manifold sunken-gap.
Into an Oblivion more devoid than the seeming redeeming:
-Absolute-*Heat* to absolute *frozen...*
*-Macro- to micro-*again in permanence and all around…
-Riding: The *Reverse*-Wave...

Trailing repeated icy-moted starry-glittery scattering
Watery-flowing coalescence: boiling-down
Brightness-defusing reflection
Fading in-wake…
In the *twinkle* of stars
In the blink of an eye
Blank-lidded unsupported and fallen unremembered immemorial:
-Universal stellar-*dying.*
Drawing-back thermal-energy densely-spiral staring
Dome roofed-walls small heavy-dense iron tin-roofed
Inner-sanctum:
-Home.
-Again.

Turning-inward flattening inward-floored deforming

Strained and stressed straddling toroidal torus saddling ringing…
The Whole *deflating* at each off-Centre
Uneven energy-mass dark disc-balling
Ballooning then collapsing:
-Iron-stellar *sunken*-core…

Heavy-elemental sinking:
-Back into The Centre of The Universe starting-point…
Surfacing gaseous only *bright* hot-accretion disc burning-off:
-Universal Dark-Star...
-Universal Black-Hole! Centre of The Universe!
-The-Galaxy! EarthCentre?! Iron-carbonate Crystal-Ball...
-The Future is Now!
-The Edge of The Universe…
-The-Galaxy! EarthCentre!
-Galactic Black-Hole! Stabilising-fused clear tri-parted Helium...
-Triple *neutral*-hydrogen Proton-electron-positron…Tritium…
Gas-*cloud* swirling-decaying neutron-*neutrino*...neutralino-*fired:*
-Zero-energy *neutronium*-chamber curling-away dephasing-vacuum *disappearing*:
-Three-dimensional hollow *vacuum*-Field…
-Atomic anti-*electron* clouded…

Unapproachable *finally* seemingly stationary everything at once shadowed and *closing-in*…
No-*light* escaping absorbent spiraling inward and *disappearing*…
Surrounding barred de-clining-forcefield
Trapped dropping-into blank-gaping
Gapping a dense stilled circularity
Close to *nothingness*…

-Each One of Us:
-Singular-singularities *surviving*…
Spiderous-netted trying to climb-out return to *almost*-certainty:
-The Eternal-*infinite* certainty-of-*nothingness*…existing or not.
-Dead-stellar region massively *empty*-drained…
Fountaining-in centre...
-*White*-star…returning black-star…black-hole…white…
-Within the dense-transparency and dark-radiation…
-Of Everything…
-Each of Us a bare-*electron* voiding into *nothingness*
At the elliptic centring-edges drawn-in anomalous speeded-up
And as if being strangled breathless *snuffing*-out.

27.
As *nothing* compared to the nothingness all around; within and without *ghostly* transcendental *spooky*-distance
Leapt over-lapping under-taking
Conjoining cohering and dis-cordant…cordant…*stringing*-along…

Cacophonous de-coherent chaotic-disarray…
With a gasp a sharp intake of breath.
Totally-absorbing densely-compressed
And now exhaling exploding once-more emitting darkly-conic:
-Super-gravitational…
-*Light* and sounding electro-magnetic nuclear-crushing…
Trumpeted ear-drums beating re-*pulsing*:
-The Universe: planetary-lunar Super-Solar…
-Hyper-Galactic re-*centring*...
-Totally un-balancing eventuality…
-Massively re-heating Hyper-Universal...

-Super-Hyper Nova…
-Starburst!...
Re-forming again outward and anew:
-Hyper Supernoval-Galaxies...*stars*...
Into the still absolute uncharted cold-lighted lighting
Foundry formed-gaseous mineral-metal shooting-gallery:
-Galactic-Sun star now de-birthing...
-Iron-extruding nickel-cobalt blue-light reflecting...
-Cuprous-orange green silvering brassy zinc-alloy
sealing transitional...
-Post-transition metal Gallium carbon-group
Germanium...
-Non-living matter organic non-organic matter and
anti-matter...
-Between silicon and foil-shining halogenic red-brown
stinking corrosive biochemically poisonous deadly:
-Arsenic purplish red grey solid Silenium...
-Toxic-*vapourous* fire-retardant Bromine...
-Fluorescent orange Kryptonite...
-Actively taken up fifth ring long-living half-life halved
and halved again...
-Rubidium *nuclear*-fallout radio-active decaying-
Strontium...
Silvery very bright-blue-lining spectral yellow-green hot-dust
multiple-image-resonant:-scanning-firing all-colour alloying:
-Inner-dropping grey-*phospherous* glowing:
-Cathode *ray*-tubed televisual:...
-Yytrium-Zirconium...cancerous granuloma multiple
image-resonating:...
-Scanning-alloy: Niobium-Molybdenum...
-Higher enzyme nitrogen fixing nuclear fissioning:
-Technetium-Ruthenium electric-contact point neutron
fluxing...
-Rhodium singling outer fifth-ring emptying...
-Four ring filling paladin-Palladium...

-*All*-knowing torment sorrow tenacity holding-on and letting-go...
-Storming and more subtle heavier and lighter...
-Newly-innocent experience of *this* conscious and individual-
existence...
-And *yet*...
-In the moment selective thought-*provoking...*
-*I*maginative-memorific decision-making...
-Re-stabilising re-habituating meaningful-activity blasting-focus...
-Re-establishing leaning this-way and that elastic-*modulating*
cellular-*valve* pumping...
-Aiming-aimed non-determining non-random elemental...
Drumming determining-randomly at-first fundament-stilted tilting:
-Wobbly-welded-together welted wilting randomly-*seemingly...*
-Apparently in: Space-Time continuity...
-Mass and *e*nergy proportionate irradiating...
-As a spiked capped ball-cycling *seasonal*-landscape...
-*S*urface rolling through the hill's and valley's of Space:...
-As an atom molecule cell organism matter or material...
-A billiard-ball game?
-Or Violincello...playing Mozart?
-Not *yet...*
-Energy-Mass *shifting gear...*
-Quantum-engine fueling electron-nucleonic state:
-Planet...or Solar-System...
-Cometary meteorite or asteroid?
-I*on conductive...*corked-corded background.

Re-cording replaying: spiked speared spread-out...fielded...catcher's gloved
Curved rivers led into sea's and curving horizontal-vertical
Beyond which nothing could be seen:
-Between Ourselves...

Each-momentary embodying *full-felt-effect*: conspiratorial
collaborative
Screaming-chaotic horrific ruly-yet-*terrifyingly* at The-Same
Time...
-Everything-Else!

Colandrical calendrical cellarly beams-of silver-grey fogging
flogging...
Gaping-darkly energetic-material:
-Ancient Andromeda!
-Promethean...Now! Triangulate! *thirded:* Universal
Hyper-
Supernoval...
-Galaxies!
-Amongst now perhaps trillions of satellite-galaxies...
-The Local-Group...
-Lone-stars and planetary-and inter-stellar planetary
objects...
-Carrying rock and icicle-cometaries into still absolute
uncharted
cold-lighted: *lightning* foundry-forming:
-Hyper-galactic re-birthing *electric*-storming:...
-E8 248...
Sheltering-stable-metal-eighteen times eighteen-filled.

Outer-rings inner-octagonal doubling-quartery:
-Binary-helium failed-stellar silver-brittle poor-
metallic...
-Catalytic-converted hydrocarbonate...fifth-ring re-
starting up...
-Up again and dropping down inward...
White light re-enameling:
-Outer valence-ring added optimal superconductive
silver *flashed!*
Photographic-nitrate alloy-refined precious:

-Cadmium metal-ore indigo blue Indium-neutrons to Tin.

Stellar and Galactic-mergers creating larger-atoms rapidly:

-Neutron-capture *r*-process water-resistant hydrophobic-tailing...

-Non-corroding metal *lustrous*: Irridium ringing Platinum...

-Malleable-Gold conducted molten-metal in-time hardy hardly changing rare charged-points retaining digital-binary singular-valence…

-Ductile and rare metal non-ionising proton and neutron nucleon and non-ionised *electron...*

-Atomic-numerical non-ionised state: Name: symbol atomic-number...

-Density-standard tipping-over over-tipping -099 Sodium to Rhubidium 1.5+ 2 sulphur-graphite core-diamond density...

-*Melting*-point: of Hydrogen n/a helium-diamond graphite steeling...

-Ironic boiling-point massing neon-nitrogen carbon-citrate ionising energies...

-Valence by atomic-number and mass whole mass natural-number super indexing:

-Mercurial liquid-metal transition-flowing...

Outer-studded fifth-ring doubling; at the triple-triplicate eighteenth

Stable full-shell and sub-shell:

-The final sixth-ring banded-layering clouded the eighty-first tripling...

-Containing Hyper-Cuboidal hologrammatic post-transition Thallium halogenenic luminous *scintillation...*

-82 stable except 43 and 61 natural-earthed: 94 to 82-94 unstable nuclei-shortlived half-life daughter...

-Isotopes UR&TH 94-118 22 artificial unstable radioactive decay...

-Averagely-related weighted-electron-ion +/- nearer closer-to natural-number atomic-unit...

-Nuclear-electron binding-energy and free-*electrons... U*nambiguous ionised electron-isope allotropic:

-Leaded brother-Bismuth…

Grey photo-chemical sulphuric phosphorescent anion-cation...

-Heaving acid base-salts...

-Decaying the final-naturally-occurring de-stabling...

-But only slightly radio-active elemental comparatively low-thermal conduction now...

- Vapourising Polonium strongly superconductive end-productive...

-In-death as Astatine black-solid: Radon everywhere!

-An-Other Noble-gas…

-Silver-grey alkaline earth-metal...

-Rock-Radium again with the destructive force of the heaviest-explosive...

-Massive supernovae nebulae star and planetary lunar...

-Asteroidal meteoric cometary-birthing propelled and propelling…

-Infinite-distanced and rounded-orbital...

-Lanthanide-disturbance re-charging radio-active primordial and natural...

-Solid rare-earth oxidal-clouding watery arced arched-rodded...

-Core-clad actinoid far-flung pieces...

-Expressing emergent ever-*heavier* slowing-impulsive convulsion...

-Trace-metallic repellant nuclear-fallout silver-streaming steaming...

-Heavy heated watery atomic radiation *fizzling*:

-Frankium-Actinium and massively heated solid Thorium...

-Atomic-*dust...*

-Protactinium enriching Uranium Neptunium Plutonium...

Thirded almost half-way twice thirded the distance...
Rushing-outward toward and into:

-Absolute-Zero fissile neutron pellet heavy-nucleus fissioning split-energy *releasing...*

-Nuclear-radiation escaped-exploding rapidly-decaying everything in its path...

-Opening *massively*-heated nucleic-neutron release *chain-reactive...*

Explosive shattering and drifting manifold drawing out into the darkened open closing *disappearing*-centre:

-Galactic Universal! *heart* lunar-solar stellar-point source thermal-pluming...

-Incandescent fruiting-flowering stem-mushrooming spread fireball...

Cascading-clouding back-drafting ice-cold nuclear-winter wind freezing-out folding falling-inward and scattering splintering centre starry *snowflake...*

Outward imploded cascaded rafting howling and fired boiling-away.
The last now *freezing*-out perfecting-point im-perfected:

-Like *pointless* Action-Heroes!
-We finite phase-locked into continuous-helices spin-linking...
-Momentarily-impossibly instantly-*entangling*...entangled...
-In Super-Position...
-Of All of Time and Nature.
-With every other-incomplete dark-materiel...
-Light-material directing dimensional...
-Never-full perhaps but never Empty or Full-*enough* ...

Conserving-rotation:
-Complex dia-magnetic danger-repulsing external-order…
-Para-Dia Magnetite-field levitating diamontine-darkening carbon-*graphite…*
-Highly-divisible...*invisible…making…*
-Ourselves visible…
-Bucky-balling divisible and thus multi-plicatory *easily...*
-Cylindrical *conic*-vortex trapped fluxing reaching peak-critical-temperature again...
-Fuel-grabbing external Other-stellar power-source:...
-As within with in-definite capacitance and resistance underlying inner-mechanical...
-Phenomenal phase-transitional Super-*fluidity...*
-Each of Us…slowing-rotating through and around every moving perfecting-point(s)...
-Im-perfect finite phase-locking continuous helices spin-linked…
-Only one-directional all in complete dark and lighted-materiel directed-dimension...
-Each potential ground-Zero-implosive explosive expanding rapidly-superheated golden-red:
-Hypernoval-Universal…
-Supernoval...Solar-star…Planetary-*lunar...*
Rising into the still-surrounding empty-Space
Almost empty-vacuum readily-clustering re-configuring
fusing fissioning replicating trace liquid-gas…

Mineral-metallic empty open warp-and-weft web-patterned
dimpling warping and wefting *nothingness*
Force-field fabric-framed:
-Oceanic-dragging...
-Universal-mass moving moved movement-*forming...*

Inter-tidal choppy-wave pelagic

Crashing deep dark-explosive crashing...
Open-star birthing glittering seething stellar nighttime sea.

Opaque strung dazzling gem be-jewelling crystalline *light*
Divergent full-length forming propositional-prepositional
And more-or-less approximate and probable-exponential
potential:
    -Quark-quantised-effect integral-selecting and
cancelling-out: of all other icy-diamond graphite-polygonal
gyrating
Reeling-reeded sparks-*streaked...*
Split sharp-shards scaling quanta-clustering conglomeration.
Congealing re –apparent gas re-brighted louring-laced
embroidering
The tightly woven=weaved loosing *darkly*-flattened force-
field.
Fabric-skirting swirling clothed torn.
Golden-blue and silver-green and purple-orange
Darkly-*energetic and naturalistic-materialistic*:
    -Universal Black-Hole!
    -Black-Body of Space...white-star *radiating...*
    -Cosmic microwave- background...stellar-
radiation...stellation...
The expansive sum of seemingly infinitesimally and
seemingly infinitely large seemingly sterile barren and hostile:
    -All other possible Stellar-Galactic-Universe...
    -As from *the*-beginning...*conceiving....*
    -Order-out-of-Kaos...
    -Born-Living-*dying...*
Galactic-pausing
Solar-pausing...only briefly out of respect...
    -Amongst all-*possible* Universe...
    -The Best! Of The *worst...*
Lighted almost-now perfecting stabilizing triple quadruple-
symmetry
Inextricably inter-twining branching-and-flowing

Meandering along dark-backwater's...

-Amongst-*all* other possible...Universal-Galactic-Solar...
Systemic-Planetary and lunar roundly-accreting:

-Asteroidal meteoric and watery mass-cometary
*colliding...*colluding coalescing in-form and function:

-Timely...

-Naturally...

-Instinctually with God-like *responsibility...*

-Accountable to-Self...

-At *least...*

-All Other!

-Doing as we are done-by...

-Only ever consciously and thus meaningfully in-retrospectively:

-The Past. Known. Unknown. The Future?

-*Unknown*...as well *that* After The-Event: The-Act: *ascribed* to-Other
as self always one step-ahead of barely-conscious *thought*:

-After The-Event *even as brilliantly Grandly-designing as-Ourselves...*

-With all-possible potential probabilities thought-out
complex-timing and motion-throughout...

-Based-on the prior-knowledge of each experimental experience-
fading into only vaguely remembered-belief:

-EarthCentres...Spaceship! EarthCentre! Lifeship! Solar-Lunar
Lightship! Cataclysmic-catalysing synergetic *fallaciously*
inevitably:

-Our doing! Naturalistically as at each-
moment...ephemeral replaced...

-Super-ceding so-quickly as to seem *smoothed-out*: In-Time. In-Truth untruth and outside-of-each filling-in-the-gaps filing-outward and inward-cascading by-degree of falling-into and out-of:

-The Universe...

Toward what we now know positively to-be:

-The Centre-of-*The*-Universe..

-At The Edge....of *this* Universe:...
-A phosphorescent six-shell...
-Oxygen-forming inside-firing...

Of many-trillionic atomic-*cloud* triangulate-particulating:
-Quark-subatomic particulation...
-Planetary-lunar asteroidal and cometary meteoric...
-Galactic-*stellar*...ferro-magnetic...
-Magnetite-appetite...
-Repellent and attractant by abundance...
-Universe-colliding...
-Inter-galactic stellar-mass diffusing-plasma thickening and
thinning-out...
-Storming solar-winds resonating-planets continuously-
forming being destroying and re-forming...
-Each-Ourselves forever imminent to the end of this our uniquely and *vibrant*-formed...stormed.

Being-formed and *still-forming* Universe...*empty-nothingness*-lit...heated-*cooling...*
-Entropic-destruction...
-With *diagnostic*-ardour...
-*This* Universe!
-OurSelves!
-EarthCentres!!

The Universe gets hotter in places colder
From Absolute Zero measurable
In other words by us and thus:
Sensible by-us through-ourselves
Or instruments that we-make and calibrate...
Universal-Hot; just got a whole lot hotter...

For Professor Stephen Hawking.

Quotations included in the text include Homer Shakespeare Shelley/Woolstencraft Butler Woolf Einstein Buckminster-Fuller Atwood; and many other words and ideas plagiarized...

Also by M.Stow:

WarFair4

Walter Mepham

ArcTol

Tan Pan-Gou

EarthCentre: The End of the Universe

The End of the Universe: Universal Verses 2

The End of the Universe: Universal Verses 3

The End of the Universe: Universal Verses 4

The End of the Universe: Universal Verses 5

The End of the Universe: Universal Verses 6

The End of the Universe: Universal Verses 7

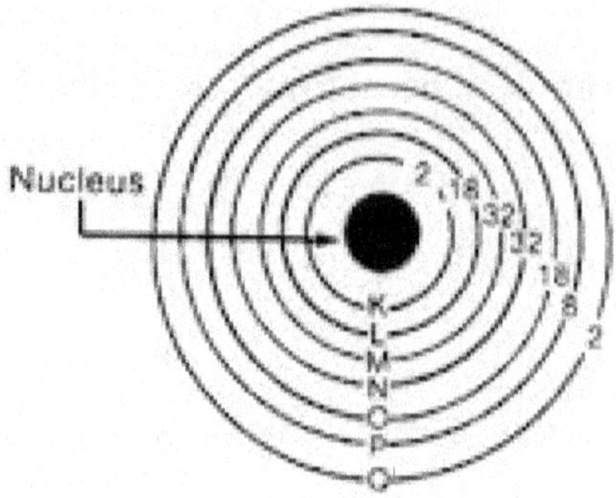

major energy level	K	L	M	N	O	P	Q
maximum number of electrons	2	8	18	32	32	18	2

Atomic nucleus and electro-magnetic cloud ring's optimum electron points of oscillation to absolute atomic number (here synthetic 112 Copernicium) The highest naturally occurring element is Uranium (92). Below: a copper atom (29) with super-conducting (in a vacuum) weakest outer-valence electron allowing field-current to pass through cable or points (aluminium silver and gold are also conductive single valance atoms).

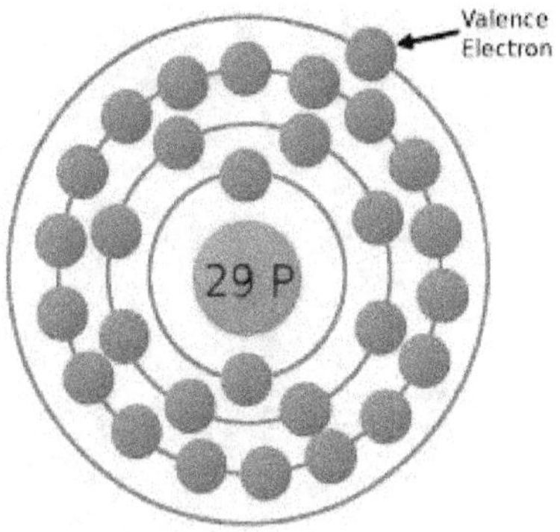

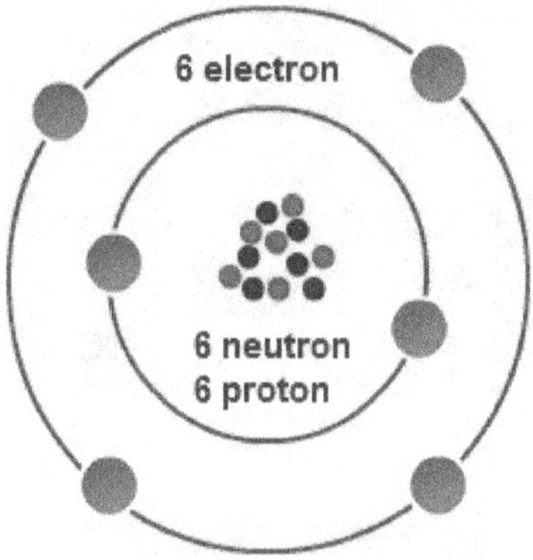

Carbon atom (above) showing stable combined neutron and proton

nucleus and an equal number of stable electrons in their rings or circuits forming an electro-magnetic protective cloud or field that surrounds the nucleus at the atomic-level (as crust and mantle is heated by the Earth's molten iron nucleus/core; as Planets and Moons surround The Sun; and Stars and Galaxies surround The Galactic Centre).

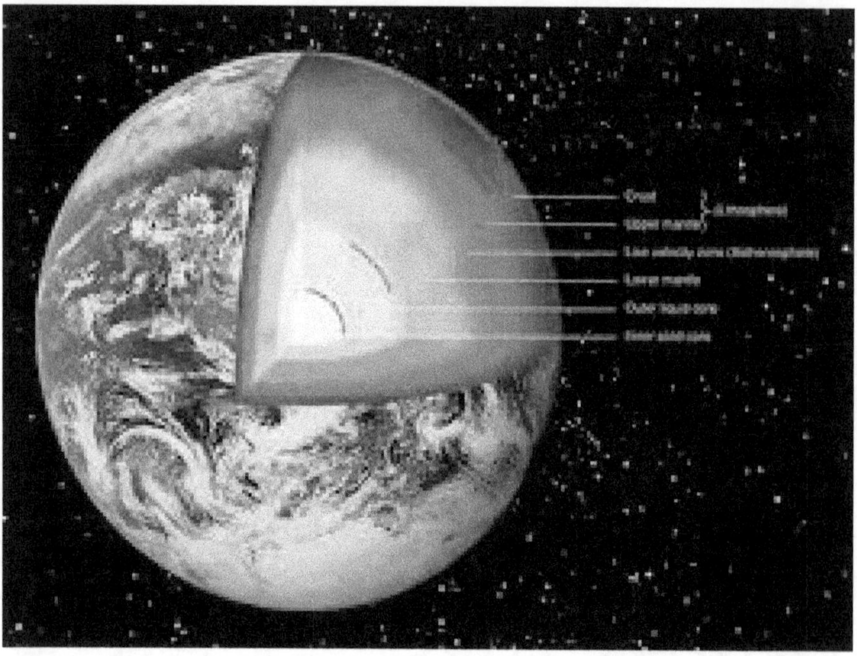

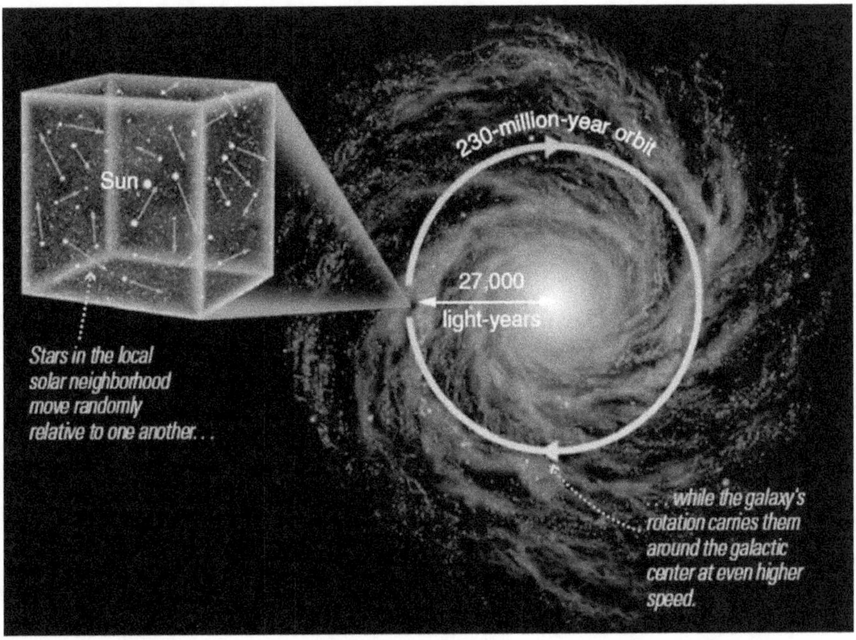
230-million-year orbit
Sun
27,000
light-years
Stars in the local solar neighborhood move randomly relative to one another. . .
. . . while the galaxy's rotation carries them around the galactic center at even higher speed.

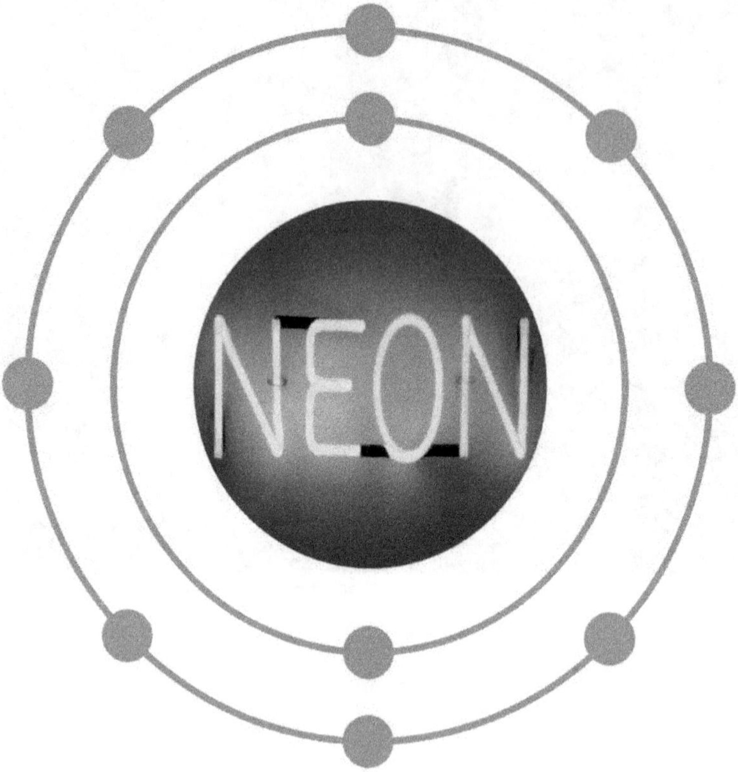

(diagrams permissions requested: chemistry.tutorcircle.com learn.sparkfun.com thinglink.com et al)

A Science fiction-faction production.

www.ingramcontent.com/pod-product-compliance
Lightning Source LLC
Chambersburg PA
CBHW051218170526
45166CB00005B/1945
* 9 7 8 1 4 9 2 7 4 6 7 7 5 *